How Einstein Ruined Physics

Motion, Symmetry, and Revolution in Science

Roger Schlafly

How Einstein Ruined Physics:
Motion, Symmetry, and Revolution in Science

Front cover art is from a 1979 USSR (CCCP) postage stamp.
It celebrates Albert Einstein, a century after his birth.

Back cover:
Einstein was the greatest public idol of the twentieth century. He became famous among the general public for being a great and humble genius, among intellectuals for discovering relativity, among philosophers for redefining the scientific revolution, and among physicists for his pursuit of a unified field theory. These are all great mistakes, and they have driven physics research into the study of alternative universes and other ideas that have no scientific foundation.

This book carefully dissects these myths. It explains the development of relativity in the context of an ancient struggle to understand motion and symmetry. It shows that Einstein's role has been exaggerated, and has been used to undermine science. Physics is no longer the science that all other sciences emulate.

Edition 1.00a.
Additional information is at www.DarkBuzz.com.

ISBN-13: 978-1461120193
ISBN-10: 1461120195

Contents

0. Introduction

Albert Einstein was my hero. A century ago, he used mathematics to synthesize a new view of space and time, and new principles of physics that have guided physics ever since. He came out of nowhere to invent relativity in a burst of abstract creativity that no one has ever seen before or since. He was the great genius who defined the modern age, including everything from lasers to atom bombs.

It is all a myth. Einstein did not invent relativity or most of the other things for which he is credited. He is mainly famous for popularizing the discoveries of others. We have all been duped.

The mathematics of relativity may be found in college textbooks, and it is not detailed here. But at the core of relativity are some basic ideas about motion and symmetry that can be traced back many centuries. One of those is the motion of the Earth. The ancient Greeks had good arguments for and against the idea that the Earth moves around the Sun. These issues were not resolved by Copernicus or Galileo. Until the invention of relativity, there were still experiments that seemed to imply that the Earth was motionless.

The Einstein myth is part of a larger set of myths about the nature of science, and about how new ideas replace old ideas. The first is the flat Earth myth. Millions of people have been taught that ancient and medieval scholars were too stupid to figure out that the Earth was round, and that the truth was suppressed by religious and other narrow-minded authorities until Christopher Columbus proved them wrong. In fact the ancient Greeks deduced that the Earth was round and there has been no serious dispute about it ever since.

The next myth concerns detecting the motion of the Earth. It is the most fundamental question in all of physics. Does the Earth move? How can such motion be demonstrated? Again, millions of people have been taught false and nonsensical stories about how ancient astronomers were unscientific or about attempts to suppress the truth. According to the theory of relativity, motion is relative, and uniform motion is undetectable except in comparison to something else.

The philosophical myth is that science advances by paradigm shifts, with the best examples being the Copernican revolution and Einstein's relativity. But the whole concept is a gross distortion of history, and a baseless attack on the value of science.

The Einstein story has been told many times, and always incorrectly. The story is used to define what science is, how scientific ideas come to be accepted, and where science is going. And these lessons are entirely fallacious. Today professors commonly teach that scientific progress is all an illusion, and that acceptance of scientific theories has more to do with fashion than objective truth. If we cannot be sure that Copernicus was correct about the motion of the Earth, they reason, then we cannot be sure about anything. We can only arrive at truth by following Einstein's example; searching for a unified theory of everything, ignoring experiments, and using philosophical arguments to convince everyone of the new paradigm. The argument is as silly as it sounds.

The discovery of relativity was indeed one of the greatest breakthroughs in the history of science, but the lessons being drawn from the Einstein myth are false, destructive, and contrary to the scientific method. Einstein is credited with relativity for reasons that have nothing to do with any objective reality, and his idolizers use his story to promote unscientific theorizing.

Today's views of science are sharply divided into two factions. The pedestrians believe in scientific progress, and that it is accomplished by the sort of observation-hypothesis-experiment methodology that is practiced in school science projects. The other view is that of the elite intellectuals who insist on heaping the greatest praise on work with no measurable or rational advantages.

This book explains the history of relativity, and how little Einstein had to do with it. It focuses on the fundamental concepts of motion and symmetry, and how these concepts have puzzled brilliant scientists for centuries. The first real science was astronomy, and the study of motion and symmetry in the sky guided all subsequent developments in physics.

If you drop a ball while you are running, will it land at your feet or behind you? You might think that the ball would land behind you, but it will land at your feet. Likewise, if you drop a ball inside a car, it will land at the point directly below the drop, whether the car is stationary or moving with uniform velocity. Also, the whole Earth could be moving at 1000 miles per hour, but you cannot notice it by dropping a ball.

Instead of dropping balls, you can do a fancier experiment with laser beams, and you still will not detect the Earth's motion. Relativity was discovered when physicists tried to reconcile such experiments with their knowledge about how light behaves like a wave in the aether. They were led to the conclusion that either the Earth was stationary, or we needed a new concept of space and time. The theory of relativity is the idea that there are symmetries of space and time that allow light to be a wave and motion to be relative.

Relativity led to a profound new understanding of motion and symmetry, and that inspired a lot of 20th century physics. It was real progress from real science. Somehow philosophers have concluded that science jumps irrationally without necessarily progressing towards truth, and physicists have been persuaded to pursue Einstein's dream of a unified theory of physics that is unguided by any observational evidence. This book explains how the Einstein myth has led to a seriously mistaken view of what science is all about.

About the author

Roger Schlafly got his BSE degree from Princeton University in Electrical Engineering (Engineering Physics program), and his Ph.D. in Mathematics from the University of California at Berkeley. He has taught at the University of Chicago and the University of California at Santa Cruz. He currently lives in Santa Cruz, California.

Joe Forshaw and Elvira Colburn read the manuscript and made many helpful suggestions.

1. Motion, symmetry, and Einstein

Relativity got its name from the idea that motion is relative. We say that a tree is at rest because it is firmly planted in the ground and not moving anywhere. We ignore the fact that the Earth is moving. The tree is motionless, relative to the Earth.

Another way to express the relativity of motion is to say that there is a symmetry between stationary and moving observers. This chapter explains motion and symmetry in order to understand Einstein and relativity.

Motion and Symmetry

The most basic concepts in science are motion and symmetry.

Physics has always been about the study of motion. The science of physics describes rolling wheels and falling rocks. It predicts the trajectory of a cannonball. It is behind the gears and the pendulum of a mechanical clock. It predicts the sunrise and the ocean tides. It explains the mechanical workings of your car, and how your car turns the non-moving energy of gasoline into motion. It is used in all the technologies that make the modern world possible.

We eat food, and our bodies are able to turn that that food into motion. How does it do that? Somehow food has energy, and your body turns that energy into motion. Where does the motion go? In the 1800s it was discovered that heat is just molecular motion. A substance feels warm because it is made of atoms and the atoms are vibrating. When you drop a rock into the ground, the motion of the rock does not just disappear; it turns into motion of the atoms in the rock and the ground. The motion turns into heat.

The ancients were inspired to do science by watching the motion of the heavens. It must have seemed as if they were all in the midst of a gigantic clockwork, even though clocks had not been invented yet. The motion of the Sun, Moon, stars, and planets seemed to impose a mathematical order on a chaotic world. By tracking the sky, they learned when to plant their crops and when to expect high tides. They could even predict spectacular events like eclipses.

The concept of symmetry is just as basic as motion.

Our bodies are symmetrical because the left side looks like the mirror image of the right side. On the outside, at least. Most other animals also have a right-left symmetry. A starfish also has a five-fold symmetry. The Moon looks like a perfect round disc when it is full. A round disc has a circular symmetry in that it can be rotated and it still looks the same.

Discovering symmetries has also been crucial to the development of science. It is related to the general problem of finding patterns in nature. Suppose I drop a rock and watch it fall. Then I pick it up, move to a second place, and drop it again. I expect it to fall again, just like it did before. The reason is that there is a symmetry in the laws of nature. The laws that apply at one place and time are the same as those that apply elsewhere.

Science is all about making observations of nature, and finding patterns in those observations. Those patterns are often the result of underlying symmetries. The more symmetries we find, the more we reduce nature to something more understandable. The history of physics is largely the history of finding symmetries in nature.

The concepts of motion and symmetry are closely related. A symmetry can be expressed as an invariance of a motion, which is just a fancy way of saying that moving something leaves some structure unchanged.

Consider a ball moving uniformly in a straight line. That is motion in its simplest form. It is also symmetry in its simplest form, as every point on the line looks like every other point on the line. The English physicist Isaac Newton's first law of motion says that an object will move uniformly in a straight line, in the absence of friction or other forces. It could have also been called his first law of symmetry. Newton's second law of motion tells how that symmetry is broken if a force is applied.

Motion and symmetry can be two opposite ways of looking at the same thing. The motion describes how the object moves, and the symmetry describes how it is limited.

Long before Newton, ancient scientists struggled with the concept of motion. The Sun and Moon appear to be in motion, but do they really move? Does the Earth move? These questions turned out to be extremely difficult. Newton and others convinced everyone of the motion of the Earth in the 1600s, but even as late as 1900 there were experiments that inexplicably failed to detect the motion of the Earth. The theory of relativity was created to resolve the matter.

Einstein is the most famous genius

Albert Einstein is widely considered the greatest genius of all time. Time magazine named him *Person of the Century* for the 20th century. A Gallup poll ranked him as the fourth most admired person of the century.[1] A Physics World poll named him the greatest physicist of all time.[2] After his death, his brain was cut up into hundreds of pieces and studied for clues to his brilliance. His name is synonymous with intelligence. The Walt Disney Company sells Baby Einstein products that millions of parents buy in order to make their babies smarter. He is on the Forbes magazine annual list of the ten top-earning dead celebrities, along with pop singer Elvis Presley and children's book author Dr. Seuss. Over 500 biographies of Einstein have been published.[3] No one else is even close to his reputation for genius.

And yet hardly anyone can say what it was that Einstein really did, other than that he created an esoteric theory called *relativity*.

Whatever relativity is, people know that it is important. Physicist (and former rock star) Brian Cox recently said:

> Relativity is the basis on which all of our understanding of modern physics rests. So without relativity, we would not understand how transistors work, how cell phones work, we wouldn't understand the universe at all without relativity. It is the foundation on which [all modern science] rests.[4]

Einstein is most commonly praised for his 1905 paper on special relativity. It has been called the most famous scientific paper in history.[5] He showed that time is the fourth dimension, it is said. Discovering the true nature of time was like Christopher Columbus discovering the New World. No subsequent exploration would be as important, and no subsequent physicist would find a more fundamental concept.

The idea of time being the fourth dimension[6] was previously expressed by H.G. Wells in a popular 1894 novel, The Time Machine:

> "Can a cube that does not last for any time at all, have a real existence?" Filby became pensive. "Clearly," the Time Traveller proceeded, "any real body must have extension in four directions: it must have Length, Breadth, Thickness, and – Duration. But through a natural infirmity of the flesh, which I will explain to you in a moment, we incline to overlook this fact. There are really four dimensions, three which we call the three planes of Space, and a fourth, Time. There is, however, a tendency to draw an unreal distinction between the former three dimensions and the latter, because it happens that our consciousness moves intermittently in one direction along the latter from the beginning to the end of our lives."

The book was made into a 1960 movie where the same concept was explained. As Wells explains, time is interesting as a fourth dimension if we have some way to change time, such as having a time machine. His time machine was fictional, of course.

A year after the novel appeared, the Dutch physicist Hendrik A. Lorentz published a paper in which he proposed the concept of *local time* in a moving object. Local time differed from the time measured by clocks at rest. Lorentz used the idea to explain some electromagnetic experiments. He thus invented the time machine that would justify calling time the fourth dimension.

The French mathematician Henri Poincare wrote papers on time in 1898 and 1900. He thought that Lorentz's idea of local time was ingenious, and said that clocks in a moving object would actually measure local time and be different from the clocks at rest.

At the time, Lorentz and Poincare were two of the leading intellectuals in Europe. Lorentz had helped create the theory of electromagnetism, along with Maxwell, Hertz, and others. He was known for what is now called the *Lorentz force law*, although Maxwell had published a similar equation. It was an equation that explains the motion of an electron in electric and magnetic fields. Electrons were discovered the next year, in 1896. This was before it was known that the properties of atoms were mostly electromagnetic, and Lorentz was at the forefront of finding electromagnetic explanations in physics. Poincare did pioneering work in topology, geometry, analysis, and what is now known as *chaos theory*. They each wrote papers on what would later be called relativity theory.

With this background, Einstein wrote his 1905 paper, *On the electrodynamics of moving bodies*. It had no references to any publications.[7] The closest it came to crediting previous work on the subject was this cryptic sentence, after mentioning examples of relative motion experiments:

> They suggest rather that, as has already been shown to the first order of small quantities, the same laws of electrodynamics and optics will be valid for all frames of reference for which the equations of mechanics hold good.

This is apparently a disguised reference to the publications of Lorentz and Poincare over the previous ten years. What followed was a formulation of special relativity that is mathematically and observationally equivalent to the previously published Lorentz-Poincare theory. It is not just approximately equal "to the first order of small quantities", but identical. Einstein has the same equations, and the same physical consequences.

Einstein's understanding of special relativity was superior to what Lorentz published ten years earlier, but it was inferior to Poincare's. On every essential part of special relativity, Poincare published the same idea years earlier, and said it better. It was Lorentz's and Poincare's work, not Einstein's, that led to time being considered the fourth dimension. The origins of the various aspects of the theory are detailed in Chapter 2.

Poincare's approach to special relativity included one of the most influential ideas in all of 20th century physics. His idea was to identify the symmetries of nature, to analyze the mathematical properties of the symmetry group, and then to formulate physical laws and equations that are invariant under that symmetry group. His approach guided not only our understanding of electrodynamics, but also development of modern laws of gravity, quantum field theory, and all of the fundamental forces of nature.

Time is the fourth dimension because there are subtle symmetries that relate space and time. The laws of physics respect those symmetries, and can be written with space and time variables on the same footing. There is no way to separate space and time without breaking those symmetries.

You don't have to take my word for Einstein's theory being mathematically identical to the Lorentz-Poincare theory. That is the conclusion of every single physicist and historian who has looked into the matter.[8] Many of those experts do argue that Einstein should be considered the inventor of special relativity because he attained a superior understanding of it somehow, but they all concede that Lorentz and Poincare had a functionally equivalent theory. Chapter 3 explains the fallacies in the arguments for crediting Einstein.

The explanations for crediting Einstein are not just factually mistaken. They do not make any sense. Furthermore, they are guided by a philosophy that regards the great scientific revolutions as irrational, and that idolizes Einstein for achieving such a revolution.

Relativity was developed from experiment, but historians have somehow decided that Einstein created the theory out of pure thought. Theoretical physics today is dominated by those hunting for another Einsteinian revolution, in the hopes that some enlightened genius will tell us how the world ought to be. This hope is severely misguided.

Einstein remains a controversial character today because of some non-scientific reasons. He was a socialist,[9] a Communist sympathizer,[10] a Zionist, a pacifist, a determinist, an egomaniac, a philanderer,[11] and a disbeliever in a personal God.[12] His life story has been told many times, and aspects of his personal life are objectionable to some. But none of that is my concern. I am not a relativity skeptic either. Relativity is at the core of our understanding of light, magnetism, electrons, gravity, and causality. There is no non-relativistic viable rival theory for any of this physics, as far as I know. I just want to clarify some popular misconceptions about how we got this theory.

Einstein is sometimes also credited with the atomic bomb. This is based on him doing two things: publishing the formula $E=mc^2$ in 1905, and signing a letter to President Franklin D. Roosevelt in 1939. Actually the formula had already been published by Poincare in 1900.

Curiously, Wells has a priority claim on the bomb also. He wrote a 1914 novel[13] where he predicted "atomic bombs", inspiring the Hungarian-American physicist Leo Szilard to invent the chain reaction concept underlying the atomic bomb about 20 years later. Neither

Einstein nor anyone else thought that a practical application of radioactivity was possible. It was Szilard who patented the nuclear chain reaction and who helped prepare the letter that Einstein signed. That letter triggered the Manhattan Project, which built the first atomic bomb.

The concept of symmetry was always fundamental to physics, but Poincare brought it to a whole new level when he made it a cornerstone of relativity theory. He used symmetry to guide the formulation of new physical hypotheses. The next couple of sections discuss symmetries in order to explain Poincare's idea.

Symmetry of the plane

Many sophisticated mathematical ideas have been applied to physics, but the idea that has influenced 20th century physics more profoundly than any other is the idea of the symmetry group.

A *symmetry* is just a transformation that leaves some structure unchanged. For example, you can reverse the letters in the phrase "PET STEP" and get the same letters. The reversal is a symmetry. Likewise, a rotation is a symmetry of a circle.

Symmetries form a *group*, meaning that every symmetry transformation has an inverse symmetry transformation, and any two symmetries can be combined to form a third symmetry. For example, a clockwise rotation by 30 degrees followed by a clockwise rotation by 45 degrees gives a clockwise rotation by 75 degrees. That 75-degree clockwise rotation has a inverse consisting of a 75-degree counterclockwise rotation.

There are three kinds of symmetries of the plane: translations, reflections, and rotations.

The symmetries of the plane are precisely those transformations that preserve the distances between points. The translations, reflections, and rotations have the property that the distance between any two points will be the same as the distance between the transformed points. And any transformation with this distance-preserving property can be written as a composition of translations, reflections, and rotations.

The formula for the distance between two points in the X-Y plane is given by the Pythagorean theorem of ordinary Euclidean geometry.

That theorem says that the square of the hypotenuse of a right triangle is equal to x^2+y^2, where x and y are the lengths of the legs (that meet at a right angle). The symmetries of the plane can be understood as those transformations that preserve this formula. For example, negating x is such a symmetry.

Thus the plane can be understood in terms of the algebraic properties of its metric structure, or the geometric properties of its symmetry group. The metric structure is the formula for distances given by the Pythagorean theorem. The symmetry group is the set of transformations. These views are equivalent because the symmetries are precisely those transformations that preserve the metric structure.

A symmetry of the plane must also act on *vectors*, as well as points. Imagine a particle at a particular point on the plane, moving with a particular velocity vector. Mathematically, it can be described as a vector at a point. A symmetry of the plane will take that point to a new point, and take the vector at the point to a vector at the new point. To mathematicians, the whole notion of a vector depends on understanding how it behaves under symmetries. A vector, like a velocity vector, is supposed to represent a particular geometrical direction, regardless of how coordinates are chosen. Understanding the plane means understandings how symmetries act on both points and vectors in the plane.

Symmetry groups can be either *continuous* or *discrete*. Translations and rotations are called continuous symmetries because those can be deformed to the identity transformation. For example, a rotation by 30 degrees can be deformed to a rotation by 29 degrees, then 28, and so on down to 0 degrees. A 0-degree rotation is the identity transformation. A reflection cannot be deformed like that. A discrete symmetry like the reflection is sometimes also called a *duality*.

The x-y plane can also be viewed as the plane of all complex numbers $x+yi$, where $i^2 = -1$. The two square roots of –1, $+i$ and $-i$, are similar, and there is a natural symmetry that swaps them. This symmetry of the plane is a reflection about the x-axis, and takes $x+yi$ to $x-yi$. It is a discrete symmetry. Applying it twice gives you back the complex number you started with.

One reason symmetry groups are so important in physics is that there is a deep relationship between symmetries and conservation laws. The German mathematician Emmy Noether published a theorem in 1918 proving that whenever there is a continuous symmetry group acting on a certain type of mechanical system, there is an associated conserved quantity. Similar relationships exist for other kinds of physical systems.

The simplest continuous symmetry group is the group of translations. These correspond to conservation of momentum. Saying that a system is invariant under a translation in the x-direction is more or less the same thing as saying that the x-component of momentum stays the same for all time. Invariance of translation in the y-direction is the same as the y-component of momentum being conserved over time. Understanding translational symmetries is profoundly related to understanding conservation of momentum.

You are probably wondering how momentum could have anything to do with plane geometry. Momentum is defined as mass times velocity, and nobody said anything about massive objects going anywhere. Nevertheless, when you construct a physical theory, whether it be Newtonian or quantum mechanical, momentum arises naturally and its conservation is related to space having a translational symmetry.

The other continuous symmetries of the plane are the rotations. These are related, via Noether's theorem, to the conservation of angular momentum. If a mechanical system has a time translation symmetry, that is, if the forces are independent of a time delay, then energy will be conserved. The principle that energy can never be created or destroyed is a consequence of the principle that yesterday's laws of physics are the same as tomorrow's. Until Noether's theorem, these conservation laws were thought to be the most fundamental principles in all of science. After her theorem, the symmetries were understood to be more fundamental.

Symmetries are also useful in modern particle physics. For example, electrons have the remarkable property that every electron is identical to every other electron. Therefore the quantum mechanics of electrons must be such that the equations are invariant under a rearrangement of the electrons.

Approximate symmetries are also useful. Protons and neutrons are similar to each other, but not identical. A proton has the mass of about 1,836 electrons, and a neutron about 1,839 electrons. A proton has a positive electric charge opposite to that of an electron, while the neutron has zero charge. Nevertheless, it is sometimes fruitful to assume that there is a symmetry between protons and neutrons.

Symmetry in higher dimensions

We live in a 4-dimensional spacetime, not a 2-dimensional flatland. There are three spatial dimensions, plus one time dimension. Until the 20th century, space and time were thought to be distinct concepts, but they were discovered to be interrelated in a way that makes them inseparable.

As in two dimensions, the symmetries of 3-dimensional space are the translations, reflections, and rotations. The rotations are a little more complicated mathematically because a rotation can be about an axis in any direction. For example, the Earth is rotating on its axis that runs from the South Pole to the North Pole (and on to the North Star). You could also imagine the Earth rotating on an axis through its center and pointing in any other direction. These rotations preserve the distance between any two points.

Rotations are also more complicated in three dimensions because they are *noncommutative*. If you combine one rotation about an axis through the origin with another rotation about a different axis through the origin, then the order of combination matters.

Schools usually just teach plane geometry because it is easier to draw figures on paper, but most of the principles extend to three dimensions. We have geometric and algebraic views of space just as in the last section. The symmetry group consists of those transformations that preserve the metric structure, just as in two dimensions.

Three dimensional space is *homogeneous*, meaning that there are symmetries that take any given point to any other given point, and *isotropic*, meaning that at any point there are symmetries that take any given direction to any other given direction. This is just a fancy way of saying that all the points look alike, and all the directions look alike. Once you have a symmetry group with these properties, then there is nothing special about the origin, or any other point, or any direction.

Electrons have angular momentum, usually called *spin*. As far as anyone knows, an electron is just a point particle with no diameter, so we don't know that it literally spins, but it has a property that acts just like spin and everybody calls it spin. Conservation of spin is directly related to invariance under rotational symmetry, and the rotational symmetry group is fundamental to understanding spin.

Electrons all have the same spin, as well as all having the same mass and charge. Only the axis of the spin can vary. You can imagine a spinning ball where the spinning speed is fixed. The ball can move around, and it always spins about some particular axis.

But electron spin has a subtlety that is hard to understand in terms of spinning balls. If you rotate an electron by 360 degrees, it is not quite the same! You have to rotate it by 720 degrees (i.e., two full revolutions) to get an electron with spin that is identical to its initial spin.

The jargon for this phenomenon is that the electron has spin ½. A spin 1 particle would behave as you would expect, with a 360-degree rotation leaving the spin invariant. But a rotation on a spin ½ particle only seems to do half of what you expect. It is not really correct to say that the axis vector of spin defines the electron spin. When you rotate a vector by 360 degrees you get what you started with. Instead the axis is a *spinor,* which is a mathematical object like the square root of a vector. Spinors were invented by the French mathematician Elie Cartan in 1913, and then used by the English physicist Paul Dirac in 1928 to formulate a relativistic theory of an electron. They were also used by the German physicist Wolfgang Pauli in 1927 to help explain electron spin. Spinors are essential to the stability of matter.

It is hard to understand how a 360-degree rotation could change anything, but here is a trick for seeing how a 360-degree rotation might be different from a 720-degree rotation. Take an ordinary belt, and hold one end in each hand. Rotate one end 720 degrees by turning it four half-twists. The belt now represents a continuous set of rotations from 0 to 720 degrees. Now swap hands, while carefully maintaining the orientation of the belt ends, and swap hands again. If you did it correctly, then the belt will untwist into its original position.

If you had only twisted the belt by 360 degrees, then you would not be able to untwist it. This exercise shows that a 720-degree rotation cannot be something different, while a 360-degree rotation might be.

Mathematically, there is a way to represent those 3-dimensional rotations as a symmetry group so that the 360-degree rotation can be distinguished from the identity transformation (and 720-degree rotation). The group is called SU(2) because it can also be represented as rotations on two complex coordinates. (SU stands for "special unitary".)

Angular momentum is the physical realization of rotational symmetry. Quantum mechanics teaches that angular momentum is quantized, and is always an integer multiple of a very small magic constant, called *Planck's constant* or *h-bar*.[14] The only exceptions are for some spinning particles, like electrons. The spin angular momentum of an election is half of Planck's constant. For short, we say the electron has spin one half.

Thus the symmetries of 3-dimensional space are the translations, reflections, and rotations, except that if you want to account for electrons and other spin ½ particles, you have to replace the rotation group with SU(2). That group acts on 3-dimensional space just like rotations. There is nothing special about the points in 3-dimensional space. It is the vectors for electron spin that are unusual because they are really spinors, not vectors. A 360-degree rotation in SU(2) acts on a spinor by multiplying it by –1. Another such rotation returns the spinor to its original value. It is very hard to visualize a spinor.

It appears that the laws of physics are invariant under these symmetries. Transforming by a reflection is like looking at something in a mirror. A right-handed man looks left-handed in a mirror, but it is hard to imagine any experiment functioning differently. But in fact there are particle physics experiments that function differently from the mirror image experiment.

It can be proved from general principles of relativity and quantum field theory that CPT is a symmetry, where C is *charge conjugation* (i.e., antiparticle reversal), P is *parity reversal* (i.e., reflection), and T is *time reversal*. This means that if you take a particle physics experiment, change the electrons to *positrons* and other particles to their antiparticles, take the mirror image, and run everything backwards in time, then the same experiment can be observed that way. This CPT symmetry is a fundamental duality between matter and antimatter. But none of these three transformations are perfect symmetries by themselves. There are obscure particle experiments that behave differently in a mirror reflection.

In an ordinary vector theory, such as any pre-20th-century mechanics, parity reversal P is a perfect symmetry. That is, the theory predicted that any experiment would look the same whether you watched it in a mirror or not. If a force vector moves a particle in a certain direction, then the mirror image force would make the particle move in the mirror image direction.

The first indication of a parity violation in nature was discovered by the French chemist and microbiologist Louis Pasteur. He discovered a chemical in 1848 that was not the same as its mirror image molecule. When the chemical was produced by living yeast, just one type was made. When he synthesized the chemical in his lab, both types were produced equally. Somehow all living beings are intrinsically right-handed on a molecular level, and can make right-handed molecules without making left-handed molecules.

A century later, it was learned that all life on Earth is indeed based on right-handed DNA molecules. The DNA molecule has the shape of a long double helix. The helix is like the threads of an ordinary screw. By convention, almost all screws and bolts are made right-handed so that you can fasten them to nuts by turning them clockwise with a screwdriver. Turning the screw counterclockwise removes it. A DNA molecule is a right-handed helix like a right-handed screw. Many other chemicals produced by living bodies, such as proteins, enzymes, and even sugars, are also often different from their mirror images.

Around the same time as Pasteur, the French physicist Leon Foucault demonstrated another sort of parity violation. In 1851 his pendulum slowly rotated clockwise over the course of a day, showing the counterclockwise rotation of the Earth underneath. The Foucault pendulum is a popular museum exhibit to this day, as it is the most elementary and convincing proof of the rotation of the Earth. The pendulum rotates the other way in the southern hemisphere. The effect is the same one that causes hurricanes and tornadoes to always spin counter-clockwise in the northern hemisphere. The effect can also cause bathtubs to drain counter-clockwise in the northern hemisphere, although the effect is very slight and some bathtubs drain the other way because of larger effects from the geometry of the bathtub.

Pendulums and helices can be explained with vector theories. Spinor theories present the possibility of a whole new kind of handedness asymmetry. Spinors just don't have mirror image symmetry like vectors. With a spinor theory, it is possible to have a particle that always has a right-handed spin. You could think about such a particle as always spinning clockwise in the direction that it is going, like a right-handed screw that is always being screwed into imaginary bolts that it encounters.

Modern particle physics is a spinor theory. Electric forces act just the same on right-handed and left-hand electrons, but the weak force only acts on left-handed electrons and neutrinos. No one has ever seen a right-handed neutrino. The weak force also acts on right-handed positrons and anti-neutrinos.

When we combine 3-dimensional space with time, we get 4-dimensional spacetime. Each point (x,y,z,t) is really an *event*, a point in space along with a specific date and time. A path in spacetime can be interpreted as the history of a particle. That is, the path gives the spatial location of the particle for each time value. Understanding the symmetries of spacetime is crucial to 20th century physics.

All of the symmetries of space are also symmetries of spacetime. So are time translations and reflections (where time runs backwards). The trickier symmetries are those that involve both space and time. One such transformation takes a stationary observer to one moving with constant velocity. Such a transformation is called a *velocity boost*. Velocity boosts are symmetries in spacetime in the sense that two such observers experience the same laws of physics.

Einstein is famous for relativity, and the story of relativity is discovery of the symmetries of spacetime. As the next chapter explains, the velocity boosts have some strange side effects when the velocities are large.

2. The origin of special relativity

Special relativity is the theory that Einstein is credited with discovering in 1905. It is a theory of motion and symmetry. The theory is called *special* because it only dealt with uniform linear motion and electromagnetism, and a *general* theory came later that included gravity and nonlinear motion. The general theory is more advanced, but it is the special theory that fundamentally redefined space and time.

The physicist and Einstein historian Gerald Holton wrote this about the discovery of special relativity:

> To find another work that illuminates as richly the relationship among physics, mathematics, and epistemology or between experiment and theory, or a work with the same range of scientific, philosophical, and general intellectual implications, one would have to go back to Newton's Principia. The theory of relativity was a key development both in physical science itself and in the philosophy of science.[15]

Special relativity is a theory about a symmetry group that combines space and time, and about applying those symmetries to the laws of physics. In particular, magnetism is a relativistic effect caused by applying those symmetries to the equations for electricity. But Einstein never combines space and time in his famous special relativity papers, never shows how the theory can be understood purely in terms of space and time, and never shows that any laws of physics obey those symmetries. He has to assume that electricity and magnetism obey certain symmetries, because his presentation is not sufficiently powerful to demonstrate them from spacetime properties. He had to assume what others had already proved.

Lorentz published a theory of relativity in 1895 that was an electromagnetic theory, and he and Poincare improved it in subsequent years. In 1904, Poincare made a conceptual leap to relativity being a spacetime theory that could be applied to electromagnetism. He called it an "entirely new mechanics, which would be, above all, characterized by this fact, that no velocity could surpass that of light". When he published the details the next year, he compared his new theory to Lorentz's as being like the Copernican system replacing Ptolemy's. But Einstein clung to Lorentz's theory and did not make this conceptual leap until after 1908 when the German mathematician Hermann Minkowski and many others did.

Poincare's explanation of special relativity was bold, modern, rigorous, and correct. He described it in books and lectures for the general public, and in detailed technical papers that showed that he understood it much better than anyone. He showed how it could be theoretically understood in terms of the geometry of space and time, and how it could explain physics experiments that were puzzling everyone else. It was Poincare's formulation of the theory, not Einstein's, that influenced 20th century physics and that is taught in textbooks today.

The story of special relativity has been told many times, and there are no significant facts in dispute. Everyone published his papers promptly, and there is no proof that anyone independently re-invented anything. No other major scientific discovery has been so thoroughly documented. Here is an elementary explanation of the elements of the theory that are usually attributed to Einstein, and how they were really discovered.

The relativity principle

Relativity got its name from what Poincare called the *relativity principle*. It is the idea that uniform linear motion is not detectable.

The ancient Greek philosopher Aristotle wondered why we do not feel the motion of the Earth, if it moves. If he rode a horse, he felt the motion of the horse. Today the answer is obvious. You can ride on a bus, train, or airplane and not realize that you are even moving. You can feel bumps, turns, acceleration, and braking, but not steady motion.

When Newton formulated his laws of mechanics, his most basic premise was that an object at rest is just like an object in uniform linear motion. His first law was that, in the absence of forces, an object at rest stays at rest, and an object moving with constant velocity in a straight line continues in that straight line. His second law tells how forces cause the objects to accelerate.

Newton's law clarified the notion of *inertia*. That is the quality of an object that makes it resistant to changes to its motion.

Inertia is closely related to *weight*. Physicists like to use the word weight to describe the force of gravity. Weight measures how heavy an object is. The heavier it is, the more inertia it has, and the more force is needed to move it. But the concept of weight also encompasses the effect of gravity. Astronauts are said to be weightless in outer space, but they still have inertia. It still takes a force to change their velocity.

Inertia is measured by *mass*. Informally, mass and weight are used interchangeably. You might say that your weight is 200 pounds, or your mass is 200 pounds. These mean the same thing to us because an object's mass responds to the Earth's gravity in the same way everywhere on the Earth's surface. But mass is the more technically accurate way to describe the object's intrinsic inertia.

Modern ships and other vehicles often have inertial navigation systems. These have sensors that use the concept of inertia to detect changes to velocity. They cannot detect linear motion, but they can measure acceleration, and use that information to help navigate. The sensors are increasingly popular in handheld consumer electronics. These sensors feel acceleration in the same way that you feel it when you are in a car turning a corner.

By the 1800s, it was widely understood that the laws of physics are valid in any inertial frame of reference. That means that the laws may be applied by an observer at rest, or an observer in uniform linear motion. If the observer has a sensor that does not detect any acceleration, then he can apply the laws of physics just like anyone else.

Physicists began to doubt the relativity of motion in the late 1800s. Electricity seemed to be different when moving, as explained in the next section. By 1895, when most physicists had given up on the principle, Lorentz and Poincare proposed that the principle was true as a fundamental law of nature. In 1904, Poincare stated the five or six most important physics principles, and included:

> The principle of relativity, according to which the laws of physical phenomena should be the same, whether for an ob-

> server fixed, or for an observer carried along in a uniform movement of translation; so that we have not and could not have any means of discerning whether or not we are carried along in such a motion.[16]

But to maintain this principle, they had to redefine space, time, and mass. It was the start of 20th century physics.

Electromagnetic relativity

In 1861 the Scottish physicist James Clerk Maxwell published the most important set of equations in the history of science. Some of the equations had been previously discovered by others, and together the equations explained electricity and magnetism. They soon became known as *Maxwell's equations.*

The equations describe how electric charges cause electric fields, and how moving electric charges or electric currents cause magnetism. They also explain how the fields affect the charges. The charges and fields are related by some partial differential equations that can be difficult to solve because the charges and fields affect each other at the same time. As a surprise byproduct, the equations explain light and optics as well.

These equations govern most of your everyday experiences. Unless you are a nuclear physicist, the only forces you ever encounter are gravity and electromagnetic forces. At some fundamental level, these equations govern how you live and breathe and even read this book.

Maxwell's equations are difficult. The German physicist Heinrich Hertz wrote in his treatise:

> To the question "What is Maxwell's theory?" I know of no shorter or more definite answer than the following: – Maxwell's theory is Maxwell's system of equations.[17]

Maxwell's own massive two-volume 1873 *Treatise on Electricity and Magnetism* has a thousand pages of dense mathematics. Supposedly one scientist commented at the time that he was very impressed with the treatise, but he did not see what it had to do with electricity. The subject had become too mathematical for most physicists. But there is no way to understand electric and magnetic fields without understanding these equations.

Two millennia earlier, Euclid was supposed to have said, "there is no Royal Road to geometry." By this, he meant that he could not tutor the king in geometry unless the king was willing to do a lot of hard work learning mathematics. When Ptolemy, Copernicus, and Kepler published their famous astronomical models, they were written in the language of Euclidean geometry. Their books are incomprehensible without a thorough understanding of geometry.

Likewise, the relativity papers were written in the language of Maxwell's equations. The technical relativity papers by Lorentz and Poincare are unintelligible without an understanding of Maxwell's equations. It is possible to explain the basics of relativity theory without those equations, but the historical roots of the theory are mathematical and electromagnetic.

From studying the equations, Maxwell and others came to some remarkable conclusions. They discovered that electric and magnetic fields propagate through space like waves traveling at the speed of light. In fact, radio waves and ordinary visible light are examples of these waves. Radio communications were invented as a byproduct of attempts by Hertz and others to demonstrate the waves that Maxwell's equations predicted.

For centuries, it had been debated whether light consists of particles or waves. Now physicists had conclusive evidence that light was a wave, and had equations to describe the wave. But the question remained — a wave in what medium?

Water waves travel through water. They disappear when they hit the beach. Sound waves travel through air. There is no sound in outer space. But we do get light from distant stars through what appears to be empty space. How can there be a wave without a medium? The name *luminiferous aether*, or light-bearing medium, began to be used to describe whatever medium transmits light, radio, and other electromagnetic waves. (It can also be spelled *ether*. The old-fashioned British spelling is used here so that the aether will not be confused with the chemical *diethyl ether*, and because the major aether advocates spelled it that way.) The name goes back to the ancient Greeks, who used it to describe whatever fills the void between the stars and planets. It was also the rarefied air breathed by the gods.

Another puzzling aspect of Maxwell's equations was that they seemed to conflict with Newtonian physics. Two centuries earlier,

Newton had argued that his laws of physics were equally valid in stationary and moving frames of reference. If you are on a train or airplane moving at a constant velocity, then you can play catch with a ball and not even know that you are moving, unless you look out the window. Even before Newton, the Italian astronomer Galileo Galilei made similar arguments that the Earth could be moving without us even noticing it.

Maxwell's equations seemed to be different. You could tell whether an electric charge was moving by whether it caused a magnetic field. Maxwell thought that this effect could be used to detect the motion of the Earth. His experiments failed. He wrote in his famous treatise that the induced current in a conductor depended only on the relative motion of the magnet and conductor.[18] The funny thing was that no one else could figure out an electromagnetic experiment that demonstrates for sure whether something was moving. The 1887 Michelson-Morley experiment tried to use optics to detect the motion of the Earth through the aether, but it also failed to detect any such motion.

The Michelson-Morley experiment was the culmination in Cleveland, Ohio of several experiments by the American physicist Albert Michelson. The experiment was to split a light beam into two beams, reflect them off some mirrors, and then combine them to show an interference pattern. Studying the pattern would show whether the two beams traveled at different speeds getting from the source to the destination. It was widely believed that the motion of the Earth through the aether would show changes in the speed of light, and hence in the interference pattern when the experiment was rotated. No such changes were seen, and Michelson-Morley is sometimes called the world's most famous failed experiment. The speed of light appeared to be constant in all directions.

The experiment was surprising because most physicists had abandoned the Newtonian idea that motion was undetectable. Maybe a more sensitive experiment would detect the motion, or maybe the Earth dragged the aether along as it moved through space.

The secret to relativity was hidden in Maxwell's equations. The Dutch physicist Hendrik Lorentz was a leading expert on electromagnetism, and his Lorentz force law was often considered part of Maxwell's equations. In 1895 he had figured out that the equations could be partially reconciled with a uniformly moving frame like the Earth if you are willing to modify your notions of space and time. Poincare later called these the *Lorentz transformations,* and said that someone in a moving frame might measure distance and time differently, relative to some other frame. Einstein wrote in 1935 of "the Lorentz transformation, the real basis of the special relativity theory".[19] The key to resolving the paradox was to gain a better understanding of space and time.

In 1902, Poincare published a popular book called *Science and Hypothesis* in which he boldly declared that uniform motion was undetectable. He said that all the laws of physics were valid in any such frame of reference, and that there was nothing that anyone could do to distinguish one such frame from another. He called it the *Principle of Relativity.*

The theory of relativity got its name from this principle. Poincare believed in it when few physicists did, and he stated it as early as 1895. Lorentz believed it as a low velocity approximation in 1895, and for all velocities in 1904. For Lorentz, it meant that there was a way to interpret electromagnetic variables so that Maxwell's equations were valid in any frame. For Poincare, it meant that all laws of physics were mathematically valid in all frames, so that there is no way that anyone could tell the difference. Einstein's famous 1905 relativity paper stated the principle with terminology similar to what Poincare had written, and applied it similarly to what Lorentz had done.

Poincare wrote this version of the relativity principle in 1895:

> Experiment has revealed a group of phenomena that can be summarized as follows: It is impossible to detect the absolute movement of matter, or better, the relative movement of ponderable matter in relation to the aether; all that one can find evidence of is the movement of ponderable matter in relation to ponderable matter.[20]

Ponderable matter is an old fashioned term for ordinary substances with weight, like rocks, water, and even air, but not unknown and seemingly immaterial substances like heat energy, fire, light, and electricity. The term is no longer useful now that these substances are better understood.

While Lorentz had a theory that could explain electromagnetic relativity at low velocity, Poincare argued that the Michelson-Morley experiment shows that this relativity principle holds without restriction.

Lorentz then reconciled Maxwell's equations with the relativity principle for all velocities in 1904.

Einstein first stated the principle slightly differently. He said that he was raising it from a *conjecture* to a *postulate*, and applying it to the laws of electrodynamics and optics. A conjecture is a statement that is believed, but might be proved true or false. A postulate is accepted as true, usually because it is self-evident. The terms are functionally similar in this case, but Einstein would later be called a great genius for calling the relativity principle a postulate instead of a conjecture, because he was said to be accepting the principle without regard for experimental evidence.

Einstein applied the relativity principle to the measurement of rigid objects. A stationary stick being measured by a stationary observer has the same length as when that same stick is moving and being measured by a similarly moving observer. So measurement of length and time in one frame is just like other frames. But if both frames observe a light beam and measure that beam as having the same constant speed, then each frame will necessarily see a moving stick as being shorter. This contraction was discovered as the logical consequence of the Michelson-Morley experiment, if the Earth is moving through the aether.

Poincare applied the relativity principle to all the laws of physics, including mechanics, electromagnetism, and gravity. Uniform linear motion was impossible to detect by any means. For him, the physics of one frame is deducible from the physics of another frame, if one is moving with constant velocity with respect to the other. Poincare's principle of relativity was the one that was adopted, and is the one that is in relativity textbooks today.[21]

Relativity of time

Relativity teaches that time is the fourth dimension. That does not mean that time is just another physical observable, like distance, temperature, or voltage. It means that the measurement of time depends on the observer, which is related to the way distance depends on the observer. You can try to separate space and time, but a moving observer will separate them differently. Put simply, there is no absolute time.

There is no absolute space in relativity either, but somehow that concept seems easier to grasp. People in Australia might have a different notion of which direction is up, but it is harder to understand that they might have a different notion of time.

If you and I have yardsticks, then there is an easy procedure for testing whether they have the same length. I can take my yardstick over to you, line up the yardsticks next to each other, hold them together, and compare the lengths. As long as we assume that the movement does not cause any permanent change to the length of the yardstick, any two yardsticks can be compared in this way.

Comparing clocks is trickier. If you and I have clocks, we could move them close together, hold them both still, and then test whether they show the same time and that they tick seconds at the same rate. The problem with this method is that moving a clock at close to the speed of light slows it down. So even if two clocks are synchronized, moving them around could cause them to lose that synchronization. The motion does not have any permanent effect on the rate of ticking seconds, but it would cause one clock to fall behind the other clock.

In practice, synchronicity of clocks is not a big problem, because the slow-down is only noticeable if a clock is going very fast. But it shows how tricky time can be. It is not even possible to synchronize a moving clock with a stationary clock because they do not even tick at the same rate. If two clocks are both stationary (in some frame of reference) then they can be synchronized by slowly moving them together and comparing them.

If a train is going 60 mph, or 1 mile per minute, then you can measure the speed of the train by watching it go a mile, and timing it to see if it took one minute. If you are on the train, then it doesn't seem like the train is going anywhere, but as you look through the window you see the countryside going by at 60 mph. If you are driving a car alongside the train, then the train might seem to be going another speed, depending on the speed of the car.

Light is not like the train. It always seems to be going the same speed, whether you are standing, or on a rocket ship towards or away from the Sun. But this seems to violate the principle of relativity, which says that the laws of physics should be the same, whether you are at rest, on a train, or on a (non-accelerating) rocket ship. Here is why.

If you have two very accurate synchronized clocks, then you can measure the speed of light by dividing the distance between the clocks by the time it takes a light beam to travel from one to the other. But if both clocks are on a moving train, then the distance traveled by the light beam will be different for the observer on the train and the observer at rest. The observer at rest will say that the distance traveled is not just the distance between the clocks, because the destination clock moved while the light beam going there. The result is that the two observers will see the light beam go two different distances in the same time, apparently contradicting the notion that light always goes at the same speed.

The paradox here is that measuring velocity of a material object always depends on the velocity of the observer. The peculiarity of light is that its measured speed does not depend on the observer at all.

Things get more complicated if one clock is on the train, and one is at rest. It turns out that two clocks can be synchronized if they are not moving with respect to each other, but a moving clock cannot be synchronized to a stationary clock. Motion distorts measurement of distance and time. The paradox is resolved by saying that the moving observer sees the light travel a different distance and a different time, and when he divides the distance by the time to get the speed, he gets the same speed of light that everybody else does.

Poincare started to address these issues in an 1898 philosophy article on the measure of time. He stressed the difficulty in deciding whether distant events were simultaneous. He declares that it is a postulate that the speed of light is a constant. To him, the speed of light was not just some observational fact. It was a convention for how we relate space and time.

If everything in the universe suddenly got twice as big, we might not notice because our yardsticks would also be twice as big. Likewise if time were somehow speeding up or slowing down, we would not notice that either because our clocks would be behaving the same way. Our units for measuring distance and time are chosen as a matter of convention. We could not detect a change in the (vacuum) speed of light either, as the measured speed is just an artifact of how our yardsticks and clocks are related.

Poincare followed up with a more mathematical article in 1900 where he further explained clock synchronization, local time, and Maxwell's equations. He starts by apologizing for having criticized the Lorentz relativity theory, and said that a good theory is flexible and adaptable to objections. Lorentz wrote a brilliant paper in 1895 where he invented a concept he called *local time* to help explain electromagnetism. He was trying to explain why certain electromagnetic effects were the same in different moving frames, in apparent contradiction to Maxwell's equations. He showed that Maxwell's equations would still be approximately valid in the moving frame, if only he used local time and other transformed variables in the moving frame. Thus each moving frame had its own concept of local time that might disagree with the local time of other moving frames. In other words, Lorentz's local time was relative.

Poincare described a clever method for synchronizing two stationary clocks. The clocks send light signals to an observer, who then makes allowances for the time that light takes to travel the distances from the clocks. The method does not require any low velocity approximations, as the clocks are never moved. It also works just as well if both clocks are moving with the same velocity. Thus time can be defined throughout any frame of reference, just as distance can be.

Poincare showed that two clocks could be synchronized if they are not moving relative to each other. If the clocks are nearby, then they are synchronized if an observer sees the same time on both clocks. If the clocks are far apart, then the observer has to make an adjustment to allow for the fact that it takes some time for a light signal to get from the clock to the observer. Two clocks are synchronized if the observer sees the same time, after making the adjustment.

Poincare's method does not actually require knowing the distance between the clocks. Each clock can broadcast its own time. Even if the clocks were already synchronized, it will look to an observer at one clock that the other clock needs a forward adjustment, because of the transmission time of the signal. The clocks are considered synchronized if they appear to each other as needing the same forward adjustment. Otherwise, one clock may be adjusted so that they are synchronized. By synchronizing clocks, a frame of reference can have a consistent standard of time throughout the frame.

Poincare realized that a clock in a moving frame would have to show local time. Local time is really just time, in that moving frame. Time is "local" in the sense that it is specific to the local frame and might differ from other moving frames. He calls it local time as a way of crediting Lorentz, even though Lorentz might not have realized that clocks would show local time. Here is how Poincare describes synchronizing clocks in 1904:

> The most ingenious idea has been that of local time. Imagine two observers who wish to adjust their watches by optical signals; they exchange signals, ... The watches adjusted in that manner do not mark, therefore, the true time; they mark what one may call the local time, so that one of them goes slow on the other.

That is, each observer sees the other one as moving, and as having a watch running more slowly than his own watch. This is a paradox; it appears to be a contradiction. It is easy to imagine one clock running slower than another, but much harder to imagine two clocks appearing to run slower than each other. Poincare understood that the situation is symmetrical, and neither clock is any more correct than the other. To each one, the moving clock appears to run slowly, and there is no contradiction.

Some epistemologists are troubled by statements like Poincare saying that watches do not mark the true time, because there is no observable true time. It is like saying that the golden mountain does not exist. For some philosophers, this is a sensible statement, but for others, it is confusing because they say that we cannot speak of the golden mountain at all if it does not exist. Poincare did not know that cosmologists would later figure out how to define true time in terms of the origin of the universe, and thereby estimate its age. He was just saying that observers see local time on their local watches, and that local time is not necessarily the same as whatever global time might be defined. Lorentz and Poincare are often accused of believing in a unique true time that is defined by clocks at rest in the aether, but they actually said no such thing and the concept plays no role in their theories.

Einstein said in 1907 that he regarded relativistic time as the critical idea:

> All that was needed was the insight that an auxiliary quantity introduced by H.A. Lorentz and denoted by him as "local time" can be defined as "time", pure and simple.[22]

What Einstein meant by this was that his theory was the same as Lorentz's except that Lorentz failed to give an operational definition of local time in terms of clock synchronization, as Poincare and Einstein had done. Lorentz first mentioned relativistic time in an 1892 paper:

> It was noted by Maxwell, that if the aether remains at rest, then the motion of earth must have an influence on the time that was required by light to travel forth and back between two points regarded as fixed to earth.[23]

From that, the constant speed of light, and the Michelson-Morley experiment, he proposed that motion affects the measurement of distance. In 1895, he proposed formulas for how the motion of the Earth influences time itself, and called it local time. Lorentz is accused of missing the insight that local time is measured by local clocks, but Poincare certainly had it, and published it in 1900. Poincare's relativity papers only talk about one kind of time, and that is the ordinary time measured by observers on ordinary clocks. He only mentions "local time" when he credits Lorentz for the concept.

Harvard science historian Peter L. Galison traces Poincare's ideas to his work for the French Bureau of Longitude in helping to develop networks for time distribution along railway lines. He defends the boldness and originality of Poincare's ideas:

> Now Poincaré is often depicted as the reactionary who was too backward to absorb fully the radical thoughts of Einstein. That, I believe, is absolutely the wrong way of thinking about it. Both Einstein and Poincaré were concerned with a new and modern physics and a new and modern world. Poincaré wrote essays and gave many lectures about the new mechanics, always emphasizing the enormous novelty of these changes in physics. It simply is not possible to describe him as simply trying to conserve, to reinstate an older physics.[24]

Since Poincare's work on the relativity of time was five years ahead of Einstein's and was well-known throughout Europe, it is a wonder that Einstein gets any credit at all. Galison gives the most common explanation:

> But Poincaré kept the fundamental distinctions between "true time" (in the frame of the ether) and "apparent time" as measured in any other frame of reference. ... "Apparent time" and "true time" were terms [Einstein] would never utter.

Galison gives an example of this supposedly non-Einsteinian usage of terms:

> Just a few months later, in the winter semester of 1906-07, Poincaré spelled out for his students precisely how Lorentz's improved "local" time fit with the Lorentz contraction to make it fully impossible to detect motion of the earth with respect to the ether. Again in 1908 he insisted that the apparent time of transmission is proportional to the apparent distance: "it is impossible to escape the impression that the principle of relativity is a general law of Nature, that one could never by any imaginable means, have evidence of anything but the relative speeds of objects" – motion with respect to the ether would never be found.[25]

When Poincare uses the word "apparent", he uses it to emphasize that he is referring to the measurements by the observer. For example, measurements on the Earth will not detect motion of the Earth. Literally, an apparent measurement means a measurement as it appears to the observer. Relativity teaches that different observers make different measurements. Poincare's usage is correct, and consistent with modern usage. Poincare only uttered the term "true time" to deny it.

Einstein does indeed use slightly different terminology. His famous 1905 relativity paper describes Poincare's clock synchronization method (without citing Poincare) and then says:

> It is essential to have time defined by means of stationary clocks in the stationary system, and the time now defined being appropriate to the stationary system we call it "the time of the stationary system."

In other words, some historians consider Einstein the greatest genius ever because he used the terms like "the time of the stationary system", instead of terms like "true time". In fact, there is no difference between Poincare's time and Einstein's time, either mathematically, physically, philosophically, or operationally. Poincare was just using Lorentz's terminology, while Einstein was using Poincare's definitions. Einstein's 1905 relativity paper added nothing to what Poincare had already published on the subject of time.

The speed of light

A fundamental fact about light, radio, and other electromagnetic waves is that their speed is constant. A recent satellite observation of an explosion billions of years away has confirmed this fact, and the press widely reported this as yet another confirmation that Einstein was right:

> Astronomers said the gamma-ray race was one of the most stringent tests yet of a bedrock principle of modern physics: Einstein's proclamation in his 1905 theory of relativity that the speed of light is constant and independent of its color, or energy; its direction; or how you yourself are moving.[26]

The research was published in the British journal Nature, which announced:

> Albert Einstein's most important contribution to physics was that the speed of light in a vacuum is constant.[27]

Apparently this is one of the most important facts in all of physics, so let's look at just exactly how Einstein discovered it. He did not observe any cosmological gamma-ray bursts or do any experiments. His proclamation was contained in just a single sentence in his famous 1905 special relativity paper:

> We ... also introduce another postulate, which is only apparently irreconcilable with the former, namely, that light is always propagated in empty space with a definite velocity *c* which is independent of the state of motion of the emitting body.

Einstein did not give any references, or explain where he got this postulate. It is not hard to guess. It is known that Einstein read Poincare's popular 1902 book, *Science and Hypothesis,* and discussed it in his book club. A friend said that Einstein was very excited about it. Poincare explains related issues, and refers to his 1898 philosophical essay on the measurement of time where he says:

> When an astronomer tells me that some stellar phenomenon, which his telescope reveals to him at this moment, happened nevertheless fifty years ago, I seek his meaning, and to that end I shall ask him first how he knows it, that is, how he has measured the velocity of light.
>
> He has begun by *supposing* that light has a constant velocity, and in particular that its velocity is the same in all directions. That is a postulate without which no measurement of this velocity could be attempted. This postulate could never be verified directly by experiment; it might be contradicted by it if the results of different measurements were not concordant. We

> should think ourselves fortunate that this contradiction has not happened and that the slight discordances which may happen can be readily explained.
>
> The postulate, at all events, resembling the principle of sufficient reason, has been accepted by everybody; what I wish to emphasize is that it furnishes us with a new rule for the investigation of simultaneity, entirely different from that which we have enunciated above.

A *light-year* is the distance light travels in one year. Poincare's point is that once astronomers started measuring distances in light-years, they were tacitly assuming that the speed of light was constant. Once you measure distance in light-years, you have defined distance in terms of the speed of light. Then the constancy of the speed of light is just a postulate or a definition, and not an observational fact. If you tried to measure the speed of light, you would always find that light travels one light-year in exactly one year. That is why Poincare says that experiments could not directly confirm that the speed of light is constant. If light traveled with different speeds in different directions, or if the speed were getting faster, we would not necessarily even notice. We could only detect a problem if some distance could be measured in two different ways and give two different results. As Poincare notes, no such contradiction has ever been found.

Poincare does not explicitly say that the velocity is independent of the motion of the emitting body, but that was considered conventional wisdom as it was implied by Maxwell's equations. It was part of the reason for Maxwell deducing that light was an electromagnetic wave in 1865. People thought that light was a wave in the aether, and that the speed of light was just a property of the aether, just as the speed of sound in air is just a property of the air. The constant speed of light is implied by the fact that some objects like Jupiter were known to be moving in different directions, and by Poincare's relativity principle that he defines in his 1902 book:

> The movement of any system whatever ought to obey the same laws, whether it is referred to fixed axes or to the movable axes which are implied in uniform motion in a straight line. This is the principle of relative motion; ...

The above writings were for non-mathematicians and had no equations. In other articles, Poincare goes much further. He extends Lorentz's work and shows that the equations for the propagation of light (i.e., Maxwell's equations) are independent of the velocity of the observer. The constant speed of light is a consequence of those equations. By 1905, Poincare was using the letter *c* for the speed of light in his papers, as others had done previously,[28] and even chose units making $c=1$ for convenience, as is often done in modern textbooks.

Thus Poincare made an astute observation about the speed of light, said that it was "accepted by everybody", and called it a postulate. He proves that the postulate is consistent with the equations for the propagation of light, and with the relativity principle. Years later, Einstein published the same postulate, and similarly called it a postulate, but did not cite any sources.

Poincare announced the new mechanics of relativity in 1904:

> From all these results, if they were to be confirmed, would issue a wholly new mechanics which would be characterized above all by this fact, that there could be no velocity greater than that of light, any more than a temperature below that of absolute zero.
>
> [footnote] Because bodies would oppose an increasing inertia to the causes that would tend to accelerate their motion; and when approaching the velocity of light, this inertia would become infinite.[29]

This is perhaps the single best one-sentence description of relativity. Absolute zero is the lowest possible temperature, and occurs when all mechanical and thermal energy has been extracted. It is −273 Celsius and −460 Fahrenheit. Getting colder is a physical impossibility, and so is going faster than light. The closer you get to the limit, the harder it is to get any closer. Going faster than light does not make any sense, just as a temperature below absolute zero does not. Both are mathematical and physical impossibilities.

Measuring the speed of light goes back to Galileo. He discovered a new celestial clock when he saw Jupiter's four main moons in a telescope.[30] The innermost one, Io, revolves around Jupiter every 42.5 hours. Its orbits can be measured precisely with a small telescope by watching for occultations, when Io disappears behind Jupiter.

Galileo was one of the first men to try to measure the speed of light on Earth, but his clocks were nowhere near accurate enough. He had to use his own pulse as a clock. The speed of light was measured in space before it was measured on Earth. After Galileo's death, in 1676, the Danish astronomer Ole Romer noticed anomalies in the Io orbits, and correctly deduced that they could be explained by the Earth's motion and the constant finite speed of light. His theory was controversial, as not everyone accepted the motion of the Earth.

The light from Jupiter and Io could take 40-50 minutes to get from Jupiter to Earth, depending on the distance, and we expect to see the occultations every 42.5 hours. But if the Earth is approaching Jupiter, then we see them a few minutes faster, because the light is taking less time to get to Earth. It is the same reason that sirens appear to have a higher frequency as they approach, and lower as they depart. That reason is called the *Doppler effect*, after Austrian physicist Christian Doppler who rediscovered it for music on a train and other Earthly signals in 1842.

A lot of people were surprised that the speed of light was finite. Aristotle said that light transmission was instantaneous, and the 17th century French philosopher Rene Descartes said:

> On the contrary, I would be worried that my entire Philosophy would be on the point of being completely overturned if any delay of this sort were to be perceived by the senses.[31]

The English astronomer James Bradley measured the aberration of starlight in 1725. He had much more accurate telescopes and other instruments, and he could detect stars appearing to move in the course of a year. He correctly attributed it to the motion of the Earth, relative to the stars, and the finite speed of light. He thus had an independent measurement of the speed of light consistent with the Jupiter-Io estimates, and observational evidence for the motion of the Earth. There was no longer any serious scientific dispute on either of these two points.

While Bradley showed that the Earth was moving relative to the stars, the 1887 Michelson-Morley experiment failed to show any motion relative to the aether. Poincare correctly deduced in 1895 that the aether was undetectable in that way. These properties of light led him to create special relativity.

Poincare was alone in seeing the speed of light as fundamental to spacetime, and not to just the laws of electromagnetism. It defines how a distance measurement can be related to a time measurement, and applies to all physical phenomena. Here is what he says in his long 1905 paper:

> If we were to admit the postulate of relativity, we would find the same number in the law of gravitation and the laws of electromagnetism – the speed of light – and we would find it again in all other forces of any origin whatsoever. This state of affairs may be explained in one of two ways: either everything in the universe would be of electromagnetic origin, or this aspect – shared, as it were, by all physical phenomena – would be a mere epiphenomenon, something due to our methods of measurement. How do we go about measuring? The first response will be: we transport objects considered to be invariable solids, one on top of the other. But that is no longer true in the current theory if we admit the Lorentzian contraction. In this theory, two lengths are equal, by definition, if they are traversed by light in equal times.[32]

Einstein used Poincare's 1900 method for synchronizing clocks, and defined distances by comparing to a stationary rigid measuring rod, like an ordinary yardstick. Poincare prefers not to use measuring rods, because of the possibility that they could be Lorentz contracted. Instead, since he had already defined time in terms of synchronized clocks, he defines distance in terms of time and the speed of light.[33] In this view, a moving yardstick is shorter because a moving clock is slower and light will not go as far in the time measured by the moving clock. Motion changes distance and time, while the speed of light stays constant. Poincare's method is more elegant because he only needs to assume that the speed of light is fundamental. Einstein's approach has some hidden assumptions, such as that a stationary rigid measuring rod returns to its original state after being moved, Lorentz contracted, and stopped.

For centuries, measures of distance and time were defined in terms of the Earth, and inaccurate clocks made it hard for ships to navigate at sea. Nowadays, the speed of light is defined to be exactly 299,792,458 meters per second. Technology has advanced to where time can be measured more precisely than anything else. A second in time is defined in terms of certain atomic oscillations, and the distance of a meter is defined in terms of how far light can travel in a certain fraction of a second. Poincare's theoretical definition has become reality.

Metric structure of spacetime

Poincare wrote a paper on Lorentz transformations in 1900, and he considered the question of what it means to compare two events. Suppose you have two clocks, and they are stationary with respect to each other in some chosen frame. Suppose that they both keep good time, but they may have been set incorrectly. How could the clocks be compared?

If the clocks are next to each other, you can just look at them and see if they both show noon simultaneously. But what if they are far apart? You cannot just look at the clocks because it takes time for light to travel from the clock to your eyes. A clock that is farther away will appear to have a different time.

In 1905, Poincare published an ingenious method for comparing two events in spacetime. If spacetime had a geometry obeying the Pythagorean theorem, then the event (x,y,z,t) would have a distance r from the origin $(0,0,0,0)$ if $r^2 = x^2 + y^2 + z^2 + t^2$. Instead, Poincare proposed using the *metric* $M = x^2 + y^2 + z^2 - c^2 t^2$, where c is the speed of light. He then showed that this metric is the same for all observers, whether stationary or moving.

Note the minus sign that could make the metric negative. It seems ridiculous to say the distance squared is an expression that could be negative, because a negative number does not have a real square root. Distances are supposed to be nonnegative. But this is no ordinary distance because time is involved, and it is just a way to compare two events numerically. Because the metric can be negative, it is called an *indefinite* metric, or a spacetime metric, or a metric *tensor*. It is sometimes also called a Lorentz metric or a Minkowski metric, but not a Poincare metric as that term is used to mean something else.

This metric has some odd properties. If a light ray (i.e., a photon) starts at the origin and goes in the direction of x-axis, then it reaches the event $(ct,0,0,t)$ at time t. The metric is zero for all such events. The photon could be a million miles away, and the metric is still zero. It is as if the path of a light beam always has a length of zero, and the metric is somehow measuring how something is different from light.

If a pulse of light starts at the origin, and goes in all spatial directions, then the light reaches precisely those events with a metric value of zero. The metric is a clever way of comparing events, by automatically taking into account the fact that light is the fastest way to get from one point to another, The metric is zero if a light beam can connect one event to another. Thus use of this metric is more profound than just saying that the speed of light is constant. It is using the speed of light to relate space and time, and to compare distant events. It is a generalization of the Poincare synchronization.

Poincare simplifies the metric a little bit by assuming units where $c = 1$. This can be done by measuring time in seconds, and distance in light-seconds. That is, one unit of distance is the distance light travels in one second. Poincare had earlier determined that the speed of light is constant for all observers. The metric is then $M = x^2 + y^2 + z^2 - t^2$.

While it may seem imprecise to measure distance in terms of light-seconds, that is actually the most technologically precise method we have today. The standard meter is not defined by some platinum stick in Paris, but is defined so that light travels exactly 299,792,458 meters in a vacuum. That is about the same as one foot per nanosecond, so you also get $c = 1$ if you measure distance in feet and time in nanoseconds.

If the metric is positive, then it is the square of something called *proper length*, and otherwise it is the negative square of something Minkowski called *proper time*. Thus the metric is a measure of either proper length or proper time, and any two observers will get the same numerical values for these measurements. Even though notions of absolute space and time are lost in relativity, there are absolute notions of proper length and time between events.

Here is an example of a computation that uses the metric. There is an elementary particle called the *muon* that acts like a heavy electron. Its lifetime is only 2200 nanoseconds, after which it decays into an electron and a couple of other particles. The speed of light is about one foot per nanosecond and nothing can go faster than light, so you might expect that muons could never go farther than 2200 feet. But muons were discovered on Earth in 1936, and they were being created by cosmic rays in the upper atmosphere. The muons were traveling a lot farther than 2200 feet.

The muons do not go faster than light. The paradox is explained by using proper time to measure its lifetime. A stationary muon has a lifetime of 2200 nanoseconds. Since the formula for proper time involves subtracting time and distance, a muon can last for a much longer time if it also travels a longer distance. Alternatively, you can think of the muon lasting for 2200 nanoseconds according to its own local time, and its local time slows down when it is going fast. The discovery of muons was the first direct proof that moving clocks really slow down as relativity predicts.

The same argument shows the theoretical possibility of human space travel to other stars. Our Milky Way galaxy is 100,000 light-years wide but a traveler might only age a year when making the trip. It is still impossible with today's technology, but relativity makes it more plausible, not less.

Poincare formally makes the metric look like an ordinary distance by using another trick now called *imaginary time*. If you let $T = ti$, where $i^2 = -1$, then $M = x^2 + y^2 + z^2 + T^2$. T is not really time, but imaginary time measured in *i*-seconds or *i*-nanoseconds.[34]

The advantage of this little sleight-of-hand is that it is easy to find the symmetries. Formally, the symmetries of spacetime are just the rotations. You can just take the ordinary formula for a rotation in the plane, substitute imaginary time, and it becomes the Lorentz transformation that emerged so mysteriously from Maxwell's equations. Switching an observer from being stationary observer to one going at a uniform velocity in some direction is the same as rotating from that direction into imaginary time. You can think of it as rotation by an imaginary angle. It sounds crazy, but that is the simplest way to state the symmetries of spacetime.

It is sometimes claimed that Einstein's postulates give the simplest foundation for special relativity, but it is even simpler to assume a spacetime with symmetries given by imaginary time rotations. The formula for the Lorentz transformation follows immediately from the ordinary formula for a rotation, and the rest of special relativity follows from the Lorentz transformation being a symmetry of spacetime.

Einstein's famous 1905 paper on special relativity gave a way of deriving the Lorentz transformation and relating it to Maxwell's equation, but he did not have the concept of a metric on spacetime. Even after Minkowski published a fuller explanation of the advantages of the spacetime approach, Einstein co-authored a paper in 1908 criticizing it, and called it "superfluous learnedness". Only later did Einstein realize that Poincare's metric approach to spacetime was crucial for gravity. Einstein's 1920 book on relativity describes four dimensional spacetime and the above imaginary time trick, and says:

> These inadequate remarks can give the reader only a vague notion of the important idea contributed by Minkowski. Without it the general theory of relativity, of which the fundamental ideas are developed in the following pages, would perhaps have got no farther than its long clothes.[35]

That "important idea" is actually due to Poincare, and is in his long 1905 relativity paper. That is where Minkowski got it, after learning relativity from Lorentz's papers. The metric is not just crucial for general relativity; it is the best way to explain special relativity as well. The German mathematician Hermann Weyl wrote a 1918 book on relativity, and he described the metric as the "solution ... which at one stroke overcomes all difficulties" in Lorentz's theory.[36]

The modern view of spacetime is that it is a 4-dimensional *manifold* with a metric tensor. That means that different coordinate frames are possible, as long as they are compatible with the metric. A manifold is a mathematical concept that allows different choices of coordinates. Ideally, the laws of physics are written independently of any choice of coordinates. Or a law can be written in terms of a coordinate frame, as long as it transforms properly under the symmetries. This modern view became essential for understanding modern gravity and cosmology.

The kinematics of spacetime

Einstein was sometimes asked to explain how his theory was any better than Lorentz's electrodynamics, if they both used the same formulas. While his 1907 explanation was in terms of his interpretation of local time, he later explained it this way:

> The new feature was the realization that the bearing of the Lorentz transformation transcended its connection with Maxwell's equations and was concerned with the nature of space and time in general.[37]

The publication of Einstein's complete works included this very carefully worded claim of his originality:

> Einstein was the first physicist to formulate clearly the new kinematical foundation for all of physics inherent in Lorentz's electron theory.[38]

Note that there is no claim that Einstein's theory was better than Lorentz's in any substantive way, but only that he presented the kinematic foundation more clearly. Nor is there a claim that Einstein's formulation was the clearest, as it is generally conceded that Minkowski formulated the kinematical foundation more clearly in 1908. Nevertheless, even this weak Einstein originality claim is not accurate.

The first half of Einstein's famous 1905 paper is on *kinematics,* and did not even mention electricity. Kinematics is an old-fashioned term for the study of the physics of motion. It is distinguished from *statics,* which is the study of things at rest or in equilibrium, and *dynamics,* which is the study of how forces cause motion. Collectively, all these areas might be called *mechanics.*

The point here is that while special relativity was developed in the context of understanding electricity and magnetism, the theory is really broader than that, and it applies even when electricity and magnetism are not involved. It applies whenever there is measurement in the presence of motion. And that was supposedly Einstein's key insight. However the idea was certainly not original to Einstein, as he was years behind Poincare on this point.

Harvard professor Peter Galison also identifies the kinematics as Einstein's claim to originality:

> Between 1900 and 1904 Poincaré kept his programmatic statements about simultaneity largely separate from his explorations into the details of electrodynamics. But even when Poincaré did introduce his notion of local time into his electrodynamics to insist on the conventionality of judgments of simultaneity, he did not, as Einstein did, use light signal coordination to reorganize mechanics and electrodynamics in such a way that force free analysis of space and time clearly begin before any considerations of electron deformations and molecular forces come into play.[39]

Galison is conceding that Poincare had published explanations of simultaneity, light signal coordination, relativistic kinematics, and electrodynamics before 1905, but he prefers to credit Einstein because Poincare did not organize them in the same way that Einstein did. This is an incredibly strained defense of Einstein. Galison seems to be complaining that Poincare's papers were written over several years, and directed at difference audiences. Also, Galison pointedly excludes

Poincare's 1905 papers, which did explain relativity as a consequence of the symmetry of spacetime.

Separating the kinematics from the electrodynamics does seem to be at the core of Einstein's claim of special relativity originality. A 1905 letter said that he was working on "an electrodynamics of moving bodies which employs a modification of the theory of space and time".[40] In the introduction to his 1905 paper, he says:

> The theory to be developed is based – like all electrodynamics – on the kinematics of the rigid body, since the assertions of any such theory have to do with the relationships between rigid bodies (systems of co-ordinates), clocks, and electromagnetic processes. Insufficient consideration of this circumstance lies at the root of the difficulties which the electrodynamics of moving bodies at present encounters.

Without mentioning Lorentz or Poincare, what Einstein seems to be saying here is that the point of the paper is to give an exposition of the Lorentz-Poincare theory that separates the kinematics from the electrodynamics. It is possible that he thought that the Lorentz-Poincare theory was so well known that everyone would understand he was just trying to recapitulate and clarify an existing theory.

Separating the kinematics from the electromagnetism is indeed crucial to understanding relativity. College physics classes today often teach special relativity before electromagnetism, because it can be given a conceptually and mathematically simpler description in terms of 4-dimensional geometry. Once special relativity is learned, then magnetism can be understood as just a relativistic effect.

What Einstein does do is to start with a discussion of the foundations of space and time. Since the theory predicts that motion distorts lengths and times, it is not so obvious how these notions can be standardized. Even without relativistic considerations, it was only in the 1880s that railroad scheduling caused time to be standardized over wide areas. The measure of length (the *meter* in metric units) was defined by a platinum meter stick in Paris, and the measure of time (the day) was defined by the rotation of the Earth.

The simplest way to define lengths and times is to first define a stationary frame, and to put a special meter stick and clock at the origin of that frame. This standard stick and clock can be used to calibrate other sticks and clocks by moving them and comparing them. Thus all points in the stationary frame can have equivalent meter sticks and synchronized clocks. Moving frames can be handled by appealing to the relativity principle, except that a moving clock cannot be synchronized with a stationary clock.

Moving standard meter sticks and clocks will cause relativistic distortions, unless the motion is so slow that the distortion is negligible. A stationary stick can be moved and stopped, and it will resume its previous length, but a stationary clock will lose its synchronization after being moved and stopped. The easiest way to standardize lengths and times is to just make sure all the motion is slow enough for the relativistic effects to be negligible.

Poincare proposed a clock synchronization procedure in 1900 (and again in 1904) that allows standardizing exact time in a frame, without any assumptions about motion or slowness. It was based on the clocks exchanging light signals. He noted in 1905 that the synchronized clocks can be used to define a standard length, by the distance that light travels in a specified amount of time. Thus length and time standards can all be defined from one clock and the constant speed of light. As Poincare wrote in 1905:

> How do we go about measuring? The first response will be: we transport solid objects considered to be rigid, one on top of the other. But that is no longer true in the current theory if we admit the Lorentzian contraction. In this theory, two lengths are equal, by definition, if they are traversed by light in equal times.[41]

To Poincare, the moving meter stick method did not really separate the kinematics from the electrodynamics. The rigidity of a meter stick is determined by electromagnetic forces, so the Lorentz contraction of a meter stick could be a byproduct of electromagnetic theory, or it could be that space itself is being contracted. But Poincare was proposing a fundamental redefinition of the measurement of space and time that would apply to electromagnetism and all other physical phenomena, so he preferred to define lengths with the light speed method.

Einstein's 1905 paper defines lengths using the moving meter stick method, and defines times using the Poincare synchronization method. That was the foundation of measurement in his kinematics. He then shows how the relativity principle and the constancy of the speed of light leads to the Lorentz transformation of those lengths and times. This was not too surprising, as that was how the Lorentz transformations were discovered in the first place, many years earlier. The Michelson-Morley experiment had shown that the speed of light was constant in different frames, and the Lorentz transformations were proposed as a way of explaining the resulting paradoxes. Einstein's kinematics made it clear that an ordinary moving meter stick would have a length contraction, but that had already been a part of Lorentz's relativity theory since 1892. As far back as 1889, the Irish physicist George FitzGerald proposed explaining the Michelson-Morley experiment by a length contraction of the apparatus. While relativity was discovered as part of the study of electromagnetism and light, it always had been understood as causing the length contraction of ordinary rigid objects like meter sticks.

FitzGerald wrote a short note in 1889 where he pointed out that the Michelson-Morley experiment seems to be in conflict with the motion of the Earth through the aether:

> I would suggest that almost the only hypothesis that can reconcile this opposition is that the length of material bodies changes, according as they are moving through the ether or across it, by an amount depending on the square of the ratio of their velocity to that of light. We know that electric forces are affected by the motion of the electrified bodies relative to the ether, and it seems a not improbable supposition that the molecular forces are affected by the motion, and that the size of a body alters consequently.[42]

He was saying that the length contraction is a logical consequence of the Michelson-Morley experiment finding that the speed of light is the same in different frames of reference. Lorentz used nearly identical logic in 1892.[43] He cited two theories about how the aether is dragged along with the Earth, including an 1818 theory by the French physicist Augustin-Jean Fresnel. He said that the Michelson-Morley experiment posed a "great difficulty", and deduced a length contraction as not being as crazy as it sounds:

> I have sought a long time to explain this experiment without success, and eventually I found only one way to reconcile the result with Fresnel's theory. It consists of the assumption, that the line joining two points of a solid body doesn't conserve its

> length, when it is once in motion parallel to the direction of motion of Earth, and afterwards it is brought normal [perpendicular] to it. ... Such a change in length of the arms in Michelson's first experiment, and in the size of the stone plate in the second, is really not inconceivable as it seems to me.

Just like FitzGerald, Lorentz went on to propose an electromagnetic explanation for the contraction. He admitted that "we know nothing about the nature of molecular forces", but if they are anything like electric and magnetic forces, then the size and shape of a solid body might be modified by motion. He pointed out that we cannot measure the contraction with a meter stick because the meter stick contracts as well. The cleverness of the Michelson-Morley experiment was that it used light to compare distances in two perpendicular directions.

The amount of the contraction is tiny, unless going close to the speed of light. The entire Earth only contracts an inch or so, based on its orbital velocity around the Sun. It was the cleverness of Michelson-Morley that allowed the ability to detect very slight changes in the apparent speed of light.

Lorentz learned about FitzGerald's hypothesis through others, and wrote to him in 1894, asking for a reference so he can be credited. FitzGerald responded that he submitted the paper to the American journal AAAS Science, but the journal had since gone defunct and he did not know whether it had been published or not. The journal was not the premier science journal that it is today, but apparently FitzGerald sent it to an American journal because it was a comment on an American experiment. He also said that he was delighted to have Lorentz agree with his hypothesis, as he had been promoting the idea to others, and they were laughing at him. FitzGerald and Lorentz had independently discovered one of the greatest insights in the history of physics, and they had boldly risked their reputations on a preposterous idea, and yet neither of them showed the slightest interest in being credited for priority.

The FitzGerald-Lorentz argument is not much different from Einstein's 1905 argument that the length contraction is deducible from his two postulates. A popular Einstein biography by Abraham Pais explains it this way:

> FitzGerald and Lorentz had already seen that the explanation of the Michelson-Morley experiment demanded the introduction of a new postulate, the contraction hypothesis. Their belief that

> this contraction is a dynamic effect (molecular forces in a rod in uniform motion differ from the forces in a rod at rest) was corrected by Einstein: the contraction of rods is a necessary consequence of his two postulates and is for the very first time given its proper observational meaning in the June paper.[44]

But Einstein did not deny that the contraction was a dynamic effect, as FitzGerald and Lorentz hypothesized. Einstein expressed no opinion one way or the other on that "not improbable supposition". The kinematic first half of his paper can be considered an elaboration of FitzGerald's first sentence in the above quote, while omitting the second. FitzGerald, Lorentz, and Einstein were all proposing the length contraction of rods as a necessary consequence of the speed of light being constant for all observers, and they all gave the same observational meaning to the contraction. They all made the same argument that Pais attributes to Einstein.

FitzGerald's and Lorentz's arguments did not depend on the aether, or on the contraction being a dynamic effect, or on electromagnetism. They argued that the Michelson-Morley experiment, along with the motion of the Earth, implied that the speed of light was constant for all observers. They are led to the contraction hypothesis as the only logical way this can happen, however improbable it may seem. They mentioned the aether only to reject the aether drift theories. They proposed the dynamic effect as a plausible explanation. Lorentz wrote a later letter that argued that the dynamic effect made the hypothesis less *ad hoc*.[45] His 1895 study of electrodynamics showed that the dynamic effect was consistent with Maxwell's equations. Larmor's 1900 book on *Aether and Matter* also gave a similar argument, and quoted Lorentz's 1895 paper. They all came to the conclusion that the Michelson-Morley experiment showed that the speed of light is constant for all observers without contradicting the relativity principle, that the logical consequence was a length contraction of meter sticks and measurable distances, and that a electrodynamic effect was a possible explanation. Furthermore, a deeper analysis could be had by studying the effect of the Lorentz transformations on Maxwell's equations.

In Lorentz's 1895 book, he says that he has joined the recent opinion that all material bodies are composed of electrically charged ions that are held in electrodynamic equilibrium, and that the electric properties of matter may be explainable in terms of the behavior of such ions under Maxwell's theory. That view turned out to be correct, and later quantum mechanics theory was used to explain the equilibrium. Lorentz explains that one could imagine that the motion of a solid body causes its dimensions to change. He does not say that the change is caused by an electrodynamic effect, but he shows that if the body's molecules are in equilibrium, then the change predicted by his electromagnetic theory is the same as that needed to explain the Michelson-Morley experiment.[46]

Lorentz also considered the possibility that non-electromagnetic forces were at work. Electric forces act on charged particles, but maybe there were forces between uncharged particles that also helped hold rigid bodies together. Lorentz assumed that those forces obeyed transformation laws like electromagnetic forces,[47] so that a body in equilibrium would change shape as he predicted.

The biggest problem with the electrodynamic explanation was that there was a theorem of electrostatics that no system could be in equilibrium from electric forces alone.[48] So it was not clear at the time that electromagnetic forces were really responsible for the rigidity of solid objects. It took relativity and quantum mechanics decades to show that atoms could indeed be in such equilibrium. FitzGerald and Lorentz saw the Michelson-Morley experiment as necessitating the length contraction regardless of the forces that bind rigid bodies.

Thus Einstein derived the Lorentz transformation from the speed of light being constant for all observers, just like the other derivations before him. He also interpreted the transformation in terms of measurements by meter sticks, just as the others did. The only things that made Einstein's paper more kinematic were that he failed to mention the Michelson-Morley experiment, that he failed to mention the electrodynamic effect was a possible explanation, and that he failed to cite the earlier relativity papers.

Today's relativity textbooks cleanly separate the kinematics from the electromagnetism, but Einstein did not. The title of his paper is "On the electrodynamics of moving bodies", not "On the symmetry of space and time". The point of the paper is to explain electrodynamics, not spacetime. The first step of his separation is his two postulates — the Poincare principle of relativity applied to electromagnetism, and the constancy of the speed of light. Both of these postulates are postulates about electromagnetism. Light was known since Maxwell to be just a particular form of electromagnetic radiation, so a postulate about light is a postulate about electromagnetism.

Perhaps Einstein did not view the constancy of the speed of light to be an electromagnetic assumption, because he had just written a paper questioning the (universally accepted) idea that light is an electromagnetic wave satisfying Maxwell's equations. He said that light was composed of particles (now called photons) with discrete amounts of energy. He could not have really been rejecting Maxwell's equations because half of his relativity paper was devoted to those equations. But even if he were, then another problem would arise. An elementary consequence of relativistic kinematics is that any particle with energy must also have momentum. Einstein could have boldly and correctly predicted in 1905 that photons had momentum. But he did not, until the German physicist Johannes Stark first proposed it in 1909.

Einstein's other postulate, the relativity principle, is also an electromagnetic assumption in the way that Einstein stated and used it. He assumed that Maxwell's equations hold (in the same form) for what he called the stationary system and the moving system. The postulate was the same as the culmination of Lorentz's electron theory. Einstein is sometimes credited with proving this, but Lorentz had already proved it and Einstein just assumed it as a postulate. Thus Einstein's special relativity was entirely electromagnetic in its assumptions and its conclusions. Minkowski is usually credited with separating electromagnetism from spacetime in 1908.

Einstein's next relativity paper, also in 1905, restated his assumptions as being Maxwell's equations and the relativity principle. He said, "I based that investigation on the Maxwell-Hertz equations for empty space, together with ...". The previous paper said, "in so-called empty space."[49] His equations for empty space were the same as Lorentz's equations for the aether, and they were essential for his definition of how the Lorentz transformation affects electric and magnetic fields. A footnote said that the constancy of the speed of light follows.

Einstein's famous 1905 relativity papers might have been considered a useful step towards Minkowski's 1908 theory, except that Poincare published a much cleaner separation between electromagnetism and kinematics in 1905. He showed that no electromagnetic postulates at all were needed. If you just assume a linear spacetime with a metric that uses imaginary time, you can derive the Lorentz transformations and the essence of special relativity. Minkowski used Poincare's approach, not Einstein's.

Thus the history of relativity is that it was always a kinematic theory as well as an electromagnetic theory. The preferred interpretation today is that relativity is spacetime theory, with the electrodynamics being a consequence of the kinematics. That interpretation came from Poincare.

Relativity of motion

The relativity principle means that there is a symmetry between an object A moving towards an object B, and B moving towards A. If you are on a train next to another train, and you look out the window at the other train, it can sometimes be difficult to tell whether your own train is moving or the other train is moving. That is because you observe the relative motion. The laws of mechanics are the same on either train.

The laws of electrodynamics are not the same on the train. The equations do not appear to be, anyway. A stationary electric charge makes an electric field, but not a magnetic field. A moving electric charge makes a magnetic field. But it is all an illusion, as your electronic devices will work just the same on a train.

Einstein starts his famous 1905 paper by pointing to this apparent asymmetry in Maxwell's equations:

> It is known that Maxwell's electrodynamics — as usually understood at the present time — when applied to moving bodies, leads to asymmetries which do not appear to be inherent in the phenomena. Take, for example, the reciprocal electrodynamic action of a magnet and a conductor. The observable phenomenon here depends only on the relative motion of the conductor and the magnet, whereas the customary view draws a sharp distinction between the two cases in which either the one or the other of these bodies is in motion.

The point is that there is a physical symmetry that is not apparent in Maxwell's equations. The observable electrical effects just depend on the relative motion, so you get the same physical result whether you use the frame of reference of the magnet or the conductor.

Maxwell made the same point in his 1873 treatise. The point is also discussed in the textbook where Einstein learned electromagnetism.[50] Maxwell understood that if electromagnetism really just depends on relative motion, then there ought to be a mathematical proof of that.

Otherwise, there ought to be some experiment that could detect the motion of the Earth. Nobody succeeded in doing either until Lorentz discovered local time. Poincare appears to have gotten the word *relativity* from Maxwell. Poincare said, "Maxwell was profoundly impregnated with the sense of mathematical symmetry".[51] Maxwell wrote an 1877 book on *Matter and Motion* that described "the doctrine of relativity of all physical phenomena", and said that we are only able to talk about objects being at rest because we use language that tacitly assumes that the Earth is at rest. It then said this, in favor of position and velocity relativity:

> Our primitive notion may have been that to know absolutely where we are, and in what direction we are going, are essential elements of our knowledge as conscious beings.
>
> But this notion, though undoubtedly held by many wise men in ancient times, has been gradually dispelled from the minds of students of physics.[52]

The book also said that "it is impossible to determine the absolute velocity of a body in space", and went on to describe how experiments like Foucault's pendulum can detect the rotation of the Earth. Before Maxwell died two years later at age 48, he proposed doing an experiment to see if measuring the speed of light could detect the motion of the Earth. He did not live to see Michelson test his ideas.

Experiments like Michelson-Morley failed to detect the motion of the Earth. That led Lorentz and others to study how Maxwell's equations could possibly be valid for both a stationary frame and a moving one like the Earth. The mathematics just did not work out unless Lorentz assumed that time was different in the moving frame. By assuming that each frame had its own local time, Lorentz was able to show in 1895 that the same equations could hold in different frames. Thus, if you made the proper adjustments to length and time, electromagnetism just depended on relative motion, and there is a symmetry between the different frames of reference.

Symmetry group invariance

The essence of relativity lies in the symmetries between space and time. Lorentz (and also the German physicist Woldemar Voigt and the Irish physicist Joseph Larmor) found these symmetries while studying Maxwell's equations for electromagnetism, and Poincare named them *Lorentz transformations*. The mathematical name for a complete set of symmetry transformations is a *group*, so Poincare used the term *Lorentz group* for these spacetime symmetries in 1905. Poincare was the first to understand that these transformations form a group.

The significance of a symmetry group is that when you have a group, every symmetry transformation is just like every other. In symbols, if there is a symmetry transformation from A to B, then there is one just like it from B to A. If there are symmetry transformations from A to B and from B to C, then there is one from A to C, from C to A, from C to B, etc. A, B, and C will then be all the same, at least with regard to the properties respected by the symmetries.

Lorentz appears not to have understood that the Lorentz group was a group, because he did not explicitly say that the aether frame was just like every other frame of reference. Poincare saw the Lorentz group as a symmetry group, so that if two frames A and B are related by a Lorentz transformation, then they are symmetrical. Frame A will look like it is moving to frame B, and frame B will look like it is moving to frame A. Whichever one is the aether will not have any significance. The group symmetry proves that they are the same.

Poincare recognized that the Lorentz group was a very powerful tool, when used in conjunction with the relativity principle. The Lorentz group gives the formulas for converting from one frame of reference to another. If you could figure out the laws of physics in one frame of reference, then the Lorentz group determines the laws in all the other possible frames. Since the laws of physics are supposed to be the same in all the reference frames according to the relativity principle, then the Lorentz group should preserve those laws of physics. Poincare applied this tool to electromagnetism in 1905, and proved that the Lorentz group preserves Maxwell's equations. This was no great surprise, because the Lorentz transformations were distilled out of Maxwell's equations in the first place.

Ten years earlier, Lorentz published what he called his *Theorem of corresponding states*. That was in a brilliant book in which he proved that if you change coordinates to that of a moving observer, then there is a way to transform all the variables (position, time, charge density, electric field, etc.) so that Maxwell's equations have the same mathematical form. In his terminology, there is a correspondence of the states of these variables so that Maxwell's equations can be considered true for any observer. Lorentz's proof used some low-velocity approximations, but a subsequent paper in 1904 proved it for all velocities. This helped explain how the equations could be valid regardless of the motion of the Earth.

Einstein gave a similar argument in his famous 1905 paper. He is sometimes credited with discovering that Maxwell's equations are unchanged by Lorentz transformations, but he did not. He assumed as a postulate that the laws of electromagnetism were the same in all (inertial) reference frames, and that meant that he assumed that Maxwell's equations had the same form, after the transformations that take one frame to another. In effect, Einstein assumed Lorentz's theorem of the corresponding states, and then gave an alternate presentation of the correspondence. The advantage of Einstein's approach was that he used the Poincare synchronization to show that Lorentz's variables had the physical interpretation that Lorentz implied. The disadvantage was that Einstein had to make stronger assumptions.

A paradox of relativity is that it is hard to see how the relativistic effects could be symmetrical. If frames A and B are in relative motion, then A is moving relative to observer B, and B is moving relative to observer A. Where is gets confusing is that A's yardsticks contract and clocks slow down, as seen by B, and that B's yardsticks contract and clocks slow down, as seen by A. This seems like a contradiction, and it has caused many people to disbelieve in relativity.

Consider two parallel yardsticks. Each can measure the other, and see that they have the same length. But if one yardstick is rotated a few degrees, then each yardstick will seem shorter to the other yardstick. The yardstick is not really any shorter, but it appears to be because some of its length has been shifted into another dimension.

Likewise, a moving object appears to be contracted because some of its length has been shifted into another dimension. It has been rotated into imaginary time. Any distances measured by the metric tensor are preserved. The paradox is explained by having a symmetry group that entangles space and time. The symmetry group is the mathematical proof that the transformations are consistent.

Thus one can take the view that the length contraction is just an illusion. Nothing really gets any shorter. Objects just look shorter in a different frame of reference because they have been rotated into another dimension. Describing the contraction is like saying that the Sun rises in the East. The Sun appears to rise in our Earth-bound frame, but there is also another frame where the Sun is not moving.

The group property is really what shows that special relativity needs no privileged frame of reference. The symmetries show that one frame is just like any other frame. Poincare's approach was to find the invariants of the symmetry group, and to study them. That is how he found the metric tensor. This idea was so important that some suggested that relativity theory be called *invariance theory* instead.

Poincare studied the Lorentz group in terms of its *infinitesimal generators*. The generators are operators that give an alternative way of understanding the group. These sorts of operators later became fundamental as the observables in quantum mechanics, and are essential to 20th century physics.

The use of symmetry groups in this way turned out to be one of the most powerful and useful ideas in 20th century physics. Symmetry groups not only explained the relation between space and time, but they are at the root of the modern explanations for all the particles, forces, fields, and conservation laws.

Magnetism is a relativistic effect

One of Maxwell's great achievements was to unify electricity and magnetism in one theory. Electric and magnetic fields are interrelated in his famous equations. They were more directly unified in special relativity. Magnetism is now considered a relativistic effect.

Electric forces are fairly simple. Electric charges can be positive, like protons, or negative, like electrons. Like charges repel each other, and opposite charges attract. In Maxwell's view, one charge creates an invisible electric field surrounding it, and the field exerts a force on the other charge.

Magnetic fields are somewhat similar, except that there are no magnetic charges, as far as anyone knows. A common bar magnet or compass needle as a north pole and a south pole, but no one has ever found a north pole all by itself. A moving electric charge causes a magnetic field, and a magnetic field in turn exerts a force on an electric charge. So stationary electric charges cause electric fields, and moving electric charges cause magnetic fields as well as electric fields. An electric current may have a net charge of zero, in which case it will have a magnetic field and no electric field. A magnet will normally have a net electric charge of zero, but you can think of it as having zillions of electrons buzzing around in circles causing magnetic fields, and those fields being lined up so that they add up to a substantial magnetic field.

Distinguishing between stationary and moving charges like this is obviously contrary to the relativity principle. We are all on planet Earth, and everything is moving. So a stationary electric charge in a lab is really a moving charge if we consider the motion of the Earth. So it doesn't seem to make any sense to say that stationary and moving charges generate different types of fields.

Maxwell himself noticed that if he tried to take the motion of the Earth into account, he would get different values for the electric and magnetic fields, but the actual observable forces are the same. The distinction between electric and magnetic fields is just an artifact of our frame of reference.

Einstein wrote in 1952 that explaining this paradox was a major motivation for him:

> What led me more or less directly to the special theory of relativity was the conviction that the electromotive force acting on a body in motion in a magnetic field was nothing else but an electric field.[53]

His famous 1905 relativity paper began with this well-known observation that while Maxwell's equations seem to depend on absolute motion, the observable forces only depend on relative motion. He tries to explain why only relative motion matters. Assuming Maxwell's equations and the relativity principle, he is able to show how the electric and magnetic fields in one frame are related to the fields in another frame. The relationship is the same one that Lorentz gave in 1895.

It is possible to make a stronger statement. With relativity it is possible to deduce the laws of magnetism from those of electric fields. As Weyl explained:

> The fundamental equations for moving bodies are determined by the principle of relativity if Maxwell's theory for matter at rest is taken for granted.[54]

Weyl seems to have not recognized Poincare's contributions, and attributed the "adequate mathematical formulation" of special relativity to Minkowski.[55] Minkowski showed that the electric and magnetic fields are components of an electromagnetic field tensor, and showed how the Lorentz transformation acts on that tensor. In particular, this showed how a moving electric field becomes a magnetic field. Poincare proved it slightly differently in 1905.

Poincare and Minkowski are able to explain the transformations of electromagnetic fields as consequences of the Lorentz transformations on spacetime. The key idea is that symmetries of spacetime should symmetries of physical laws, including Maxwell's equations.

People often think of relativity as being barely measurable except under extreme conditions when objects are going close to the speed of light. But magnetism is a relativistic effect, and you are using it when ever you listen to speakers or run an electric motor.

Four-dimensional geometry

The German mathematician Hermann Minkowski is famous today for this bold 1908 announcement of 4-dimensional spacetime:

> The views of space and time which I wish to lay before you have sprung from the soil of experimental physics, and therein lies their strength. They are radical. Henceforth space by itself, and time by itself, are doomed to fade away into mere shadows, and only a kind of union of the two will preserve an independent reality.[56]

He then described how special relativity could be understood in terms of 4-dimensional geometry. His paper was sensational, and was quickly published in three journals, and as a special report of the German Mathematical Society. French and Italian translations were published the next year. Physics textbooks today prefer this approach over Einstein's, and sometimes call it *Minkowski space*.

Einstein was as shocked as anyone. He continued to write relativity papers where he translated Minkowski's 4-dimensional results into 3-dimensional terminology. As Kevin Brown explains:

> Einstein's reaction to Minkowski's work was interesting. It's well known that Einstein was not immediately very appreciative of his former instructor's contribution, describing it as "superfluous learnedness", and joking that "since the mathematicians have attacked the relativity theory, I myself no longer understand it any more". He seems to have been at least partly se-

rious when he later said "The people in Goettingen [where both Minkowski and Hilbert resided] sometimes strike me not as if they wanted to help one formulate something clearly, but as if they wanted only to show us physicists how much brighter they are than we".[57]

The wonderful thing about Minkowski's work is that it became possible to learn special relativity without learning any electrodynamics at all. You just had to learn the geometry of spacetime. Instead of talking about uniform velocities, you could talk about straight lines in spacetime. Instead of talking about the speed of light, you could talk about the spacetime metric. Instead of talking about the Lorentz transformations of Maxwell's equations, you could talk about the symmetries of spacetime. And, perhaps most surprisingly, you could use vectors that had four real numbers as components.

The concept of a vector is deceptively simple. It is taught to schoolchildren, and they learn it with ease. And yet for two centuries after Newton, it was not fully appreciated. Geometrically it is just an arrow. Algebraically, a vector in three dimensions has three real numbers as components. It is the preferred way to describe velocity, force, momentum, and acceleration. Elementary mechanics textbooks teach forces by drawing *force diagrams*, in which different forces are represented by arrows and balanced using vector addition. It is hard for modern students to see how anyone ever learned mechanics without vectors. But they did. Conservation of momentum meant that the X-component of momentum was conserved, as well as the Y and Z components. If you tried to explain to a 19th century physicist that it is conceptually easier to say that the momentum vector is conserved, he would probably say that is just a fancy way of saying that the components are conserved.

Electric and magnetic fields are vectors at every point in space and time. Faraday called them *lines of force*. They have intensity (some fields are stronger than others) and direction (they push or pull in particular directions). And yet Maxwell's equations were not written as vector equations. The electric field was written out in terms of its three components, often with different letters of the alphabet. There were separate equations for the different components.

Nowadays, all of these equations are written with vectors. Newton's second law is written as force being equal to the rate of change of momentum, with both being vectors. Likewise, Maxwell's equations are now written with vectors. Vectors started to become popular among physicists with the 1901 textbook on *Vector Analysis* based on lectures by the American mathematical physicist J. Willard Gibbs.

The subtlety about vectors occurs when they are transformed. Any symmetry of space is automatically a symmetry of the vectors in that space. If you are moving with a particular velocity vector in the direction of the *X*-axis, and the *X*-axis is rotated into the *Y*-axis, then that automatically means that you will have a velocity vector in the direction of the *Y*-axis. The mathematics of manipulating vectors in a space is derived from the geometry of that space.

The tricky part of Minkowski space was to understand the vectors. Spacetime has four dimensions, so spacetime vectors have four real numbers as components. They are sometimes called *4-vectors,* to emphasize the fact that they have four components instead of three. That means that if you choose some frame of reference for some particular event, then the vector has four components that make sense for that frame. If you switch to a different frame, then the vector has the same geometrical meaning, but a transformation must be applied to the components.

The beauty of the geometrical approach to spacetime is that it conceals the messy relativity formulas for length contraction, time dilation, and relativistic mass. They just become byproducts of geometrical symmetries. So the ordinary 3-component momentum can be combined with energy to get an energy-momentum 4-vector. No new analysis is needed to figure out how formulas for how energy and momentum change with a different reference frame, because the energy-momentum 4-vector transforms geometrically just like any other 4-vector.

The 4-dimensional geometrical approach to relativity was essential to all further research. A physicist recently wrote:

> Einstein originally formulated special relativity in language that now seems clumsy, and it was mathematician Hermann Minkowski's introduction of 4-vectors and spacetime that made further progress possible.[58]

The discovery of non-Euclidean geometries was one of the great achievements of 19th century mathematics. For two millennia, geometry was based on Euclid's axioms, in which space is flat and parallel lines stay at a constant distance apart. In non-Euclidean geometry, parallel lines can converge to a point, or diverge to being farther and farther apart. The German mathematician Bernhard Riemann showed how to calculate curvatures from the metrics in such geometries in 1854. In 1872, the German mathematician Felix Klein showed that many kinds of geometries could be understood in terms of their symmetry groups, and the invariants of those groups.[59] Spacetime became just another one of those geometries with the discovery of relativity.

The first man to realize the benefits of 4-dimensional geometry was Poincare. His 1905 relativity paper combined space and time into a 4-dimensional spacetime, introduced the metric on spacetime, and geometrically interpreted the Lorentz transformations as rotations with imaginary time. He used 4-vectors for electromagnetism, in order to prove the symmetry properties of the next section. By expressing Maxwell's equations in terms of 4-vectors that transform geometrically under the Lorentz transformations, he proved that the equations in one frame are deducible from the equations in any other frame. Once formulated this way, the proof is easy, because it is just an automatic consequence of the geometric formulation.

Casual readers of Poincare's paper might be surprised that it had anything to do with geometry, because it did not have any diagrams. Euclidean geometry, as it is commonly taught in school, is meaningless without the diagrams. But it is actually quite common for modern geometry research papers to analyze problems in terms of groups and metrics, and not have any diagrams. Sometimes geometrical ideas are better expressed with equations. Poincare was a poor artist, and maybe he did not think that diagrams were necessary. Minkowski's 1908 paper had four geometrical figures, and was much more obviously stressing the geometry. He drew two of the four dimensions, using time and one of the spatial dimensions. These diagrams are quite useful for visualizing relativity.

Poincare announced in 1907 that the rest of physics could be translated into the language of 4-dimensional geometry.[60] However, he did not pursue it further, and left it for Minkowski and others. Neither did Einstein, until a couple of years later when he became convinced that it was needed for a relativistic theory of gravity. Minkowski started in 1907 where Poincare left off, and stressed that the key to relativity is 4-dimensional spacetime, imaginary time, the metric tensor, and the Lorentz group. He said that the "theorem of relativity" is the covariance of Maxwell's equations. He succeeded in popularizing the geometrical approach to relativity, and that soon became the preferred way to understand the subject.

Minkowski learned relativity from Lorentz's and Poincare's papers. He credited Poincare in 1907,[61] in the second publication to build on his 1905 papers,[62] but conspicuously avoided mentioning him in his more famous 1908 paper. The latter paper credited Lorentz for the local time of an electron, and Einstein (erroneously) for being the first to recognize clearly that local time is time. (Poincare recognized that clearly five years before Einstein.) Even worse, it presents Poincare's 4-dimensional geometry as if it were original. Minkowski died a year later, and Einstein's reputation mushroomed as if Minkowski had given an exposition of Einstein's theory.

Minkowski credited Einstein for understanding the relativity of time, but not the relativity of space. "Neither Einstein nor Lorentz rattled the concept of space", he said. There is no absolute time, and no absolute space either. Minkowski made it clear that space and time were geometrically related, and inseparable.

Minkowski stimulated a lot of interest in relativity, and Poincare's 4-dimensional geometrical approach was further developed by the German physicists Arnold Sommerfeld, Max von Laue, and others. Sommerfeld wrote in 1910 that the Lorentz-Einstein approach had become "irrelevant".[63] Laue wrote the first relativity textbook in 1911, using 4-vectors. He said that Minkowski's pronouncement about the unification of space and time was exaggerated, but a generation of physicists learned the geometrical approach to relativity from his book. Einstein said, "I myself can hardly understand Laue's book."[64] Subsequent books by Pauli, Weyl, and others all used 4-dimensional geometry. Einstein's approach was already obsolete in 1911.

Spacetime covariance

Relativity turned Maxwell's electromagnetism theory into a spacetime theory. That did not just mean that Maxwell's equations had space and time variables. It meant that the symmetries of spacetime were also symmetries of Maxwell's equations.

When Minkowski said in 1908 that space and time were no longer separate, he also had an understanding of electromagnetism that did not separate space and time. The ideas that there was no absolute space, no absolute time, and no preferred frame of reference were extended to Maxwell's equations, by showing that the equations could be formulated in a way that incorporates the spacetime symmetries.

Poincare's 1902 book said:

> 1. There is no absolute space, and we only conceive of relative motion; and yet in most cases mechanical facts are enunciated as if there is an absolute space to which they can be referred.
>
> 2. There is no absolute time. When we say that two periods are equal, the statement has no meaning, and can only acquire a meaning by a convention.
>
> 3. Not only have we no direct intuition of the equality of two periods, but we have not even direct intuition of the simultaneity of two events occurring in two different places. I have explained this in an [1898] article entitled "Mesure du Temps." [Measure of Time]
>
> 4. Finally, is not our Euclidean geometry in itself only a kind of convention of language? Mechanical facts might be enunciated with reference to a non-Euclidean space which would be less convenient but quite as legitimate as our ordinary space; the enunciation would become more complicated, but it still would be possible.
>
> Thus, absolute space, absolute time, and even geometry are not conditions which are imposed on mechanics. All these things no more existed before mechanics than the French language can be logically said to have existed before the truths which are expressed in French. We might endeavour to enunciate the fundamental law of mechanics in a language independent of all these conventions; and no doubt we should in this way get a clearer idea of those laws in themselves.[65]

Maxwell's theory enunciated the facts of electromagnetism as if there were an absolute space and an absolute time. Minkowski's reformulation enunciated those same facts for a 4-dimensional non-Euclidean spacetime, with no absolute space or time.

The Lorentz transformations are symmetries of spacetime. That means that moving frames are just like each other, as long you make the appropriate adjustments to length and time. If you have a spacetime theory that obeys these symmetries, then doing your observations and calculations in one frame of reference should be just like doing them in another frame. It turns out that Maxwell's equations do have this property.

Some aspect of electromagnetism might seem superficially different to a moving observer, but it is not because the electromagnetism is really any different. It is only because spacetime looks different to the moving observer. If an observer knows the electromagnetic variables in one frame, then he can deduce the variables in another frame from just the Lorentz transformation in space and time. That is what makes electromagnetism a spacetime theory.

To transform Maxwell's equations from one frame to another, Poincare used a concept in 1905 that Minkowski and Einstein later called *covariance*. The equations involve quantities like electric charge density and magnetic field, and covariance tells how to convert these to a different frame. Once you have the Lorentz transformations for spacetime, you can deduce the transformations on any other physical variable, as long as it is covariant. Saying that something is covariant is a way of saying that it has geometric meaning like a vector. And that means a 4-vector, in the context of relativity.

The kinetic energy of an electron is an example of a physical variable that is not covariant. In one frame, an electron might be stationary and have zero kinetic energy, and in another frame, it might be moving very fast and have a large kinetic energy. To get a covariant quantity, you have to combine energy and momentum into a 4-component vector on spacetime.

The relationship between the words "invariance" and "covariance" is a little confusing, as there are some contexts where the words are used interchangeably. If something is invariant under a symmetry group, then it is also covariant. If you want to say that a geometrical vector is independent of the coordinates used to represent the vector, then you could say that the vector is invariant. But the numerical coordinates of that vector will be different if you change the coordinate system, so some people prefer to say that the vector is covariant. Saying that a vector is covariant is just a fancy way of saying that if you know the components in one frame, then you can deduce the components in another frame by just applying the appropriate change-of-variable rules.

For example, suppose I say that "the direction North is on your right". The direction North does not really depend on how your body is situated, so you might say that North is invariant. But my description of how you can find North did depend on your body. If you turn your body, then North might be on your left, or in front of you, or behind you. If you take my statement to be a covariant statement, then that means that you have an understanding of how you can turn your body and make the corresponding adjustment to my statement so that you still know where North is relative to your body. While the details of these adjustments may be confusing, covariance is just the concept that North has a geometric meaning.

To mathematicians, the words invariant and covariant are often superfluous. They would just give geometrical definitions of concepts like vectors, and then there is no need to say that they are covariant. The covariance is implied by the geometric definition. The term is useful when someone wants to make it explicit that a geometric definition applies.

Poincare showed how to formulate Maxwell's equations in terms of covariant physical quantities, and then showed that the entire set of equations is covariant. This proved that if Maxwell's equations were true in one frame, then they are true in every other frame as well. Those physical quantities might have different numerical values in a different frame, but those differences are deducible from knowing how the frame changed. Thus he gave Maxwell's equations a 4-dimensional geometric meaning that was independent of the frame.

The tricky part of proving covariance was transforming the electromagnetic field. The electric and magnetic fields are not simple vectors that can be transformed separately under a Lorentz symmetry. A moving electric field causes a magnetic field as a relativistic effect. The best way to do this is to combine the electric and magnetic fields into one entity (now called a *tensor*), and show how it behaves under a Lorentz transformation. Then the transformation of the electromagnetic field is a mathematical consequence of the Lorentz transformation of space and time. That is how Minkowski did it. Lorentz did not know how to do this, but correctly guessed the formula for the transformed fields by analyzing Maxwell's equations. Einstein just used Lorentz's formulas.[66] Poincare did not use tensors either, but found a way to prove covariance by using 4-dimensional vectors.

To Lorentz and Einstein, a Lorentz transformation had to be defined on electromagnetic fields, as well as space and time. The Lorentz transformation was thus a transformation of space, time, and electromagnetism. Poincare showed that the electromagnetic transformation was a consequence of understanding the Lorentz group as a symmetry group of spacetime. This property is now called *Lorentz covariance* or *Poincare covariance.* And he did it in two ways, by introducing the concepts of the covariant field equation and the relativistic action. These two concepts have been two of the most important concepts in 20th century physics.

Spacetime covariance was the essence of Minkowski's theory, and was what got everyone excited about relativity. He announced this in 1907:

> The covariance of these fundamental equations,... I will call this the Theorem of Relativity; this theorem rests essentially on the form of the differential equations for the propagation of waves with the velocity of light. ... The position of affairs here is almost the same as when the Principle of Conservation of Energy was postulated in cases, where the corresponding forms of energy were unknown. Now if hereafter, we succeed in maintaining this covariance as a definite connection between pure and simple observable phenomena in moving bodies, the definite connection may be styled "the Principle of Relativity".[67]

He was saying that the covariance of certain equations can be proved as a mathematical theorem, but the real power in the principle lies in the conviction that it is applicable to all physical variables, both known and unknown.

Covariance is what links the 4-dimensional geometry to the physics. Anyone could have noticed that Maxwell's equations involve functions of space and time, and hence functions of a 4-dimensional spacetime. The meat of the 4-dimensional spacetime geometry view is saying that the spacetime symmetries are also symmetries of the physical variables. Covariance is the mathematical link and it is the essence of why electromagnetism is a relativistic theory. It is what allows the Lorentz transformations to become symmetries of all of the physical variables.

When Einstein's friend Marcel Grossmann discovered the correct relativistic equations for gravity in the solar system, his guiding principle was to search for covariant field equations. When the German mathematician David Hilbert gave his formulation of gravity, his crucial tool was relativistic action, as well as covariance. When Dirac developed his quantum theory of an electron, his biggest breakthrough was to find a Poincare covariant equation. The subsequent development of quantum field theory used a relativistic action. Poincare covariance is one of the great lessons of special relativity for physics.

Abolishing the aether

Einstein is generally credited with coming out of nowhere and revolutionizing physics by inventing special relativity from his own thought experiments. But historians who have studied the matter are startled to find that Lorentz and Poincare had every major aspect of the theory, and had published it before Einstein. But many of them credit Einstein anyway, and the main reason they give is that only Einstein rejected the aether. For example, the American mathematical physicist Freeman Dyson wrote that the only difference was the terminology used for the aether:

> Today the name of Einstein is known to almost everybody, the name of Poincaré to almost nobody. A hundred years ago the opposite was true. ... The theories discovered by Poincaré and Einstein were operationally equivalent, with identical experimental consequences, but there was one crucial difference. The difference was the use of the word "ether."[68]

The New York Times said the same thing about Poincare:

> Einstein is a household name today. But at the end of the 19th century, it was Poincaré, a mathematician, physicist, philosopher and member of national academies, who was the famous one ...
>
> Among his noteworthy feats now is what he did not do: he did not invent relativity, even though he had some of the same ideas as Einstein, often in advance, and arrived, with the Dutch physicist Hendrik Lorentz at a theory that was mathematically identical.
>
> The difference was that Poincaré refused to abandon the idea of the ether, the substance in which light waves supposedly vibrated and which presumably filled all space.[69]

Here are the major tenets of special relativity. The relativity principle that different inertial frames are indistinguishable dates back to Galileo and Newton, but it was Poincare who articulated it in the face of seemingly contrary evidence from electromagnetism. It was Poincare who coined the term "relativity" in a popular book. Different frames are related by Lorentz transformations, and the invariance of Maxwell's equations was shown approximately by Lorentz and perfected by Poincare. The speed of light is constant for all observers. Spacetime has a metric structure, independent of electromagnetism, and the symmetries of that metric structure explain the indistinguishability of different frames. There is a mass energy equivalence given by $E=mc^2$. On all of these points, Poincare was years ahead of Einstein.

But what about the aether? Should we really credit Einstein for rejecting the aether? Some say so, such as one prominent physics blogger who wrote:

> Einstein's 1905 relativity revolution can be summarized as a successful assassination attempt against the aether.[70]

Modern physicists like to make fun of 19th century physicists for believing in an aether, especially when it was described in terms that seemed to imply that it was a material substance. Suggested properties of the aether included being rigid and elastic, and having *molecular vortices,* a concept that predated the electron. But hardly anyone really believed that the aether was a material like ordinary earthly materials. The necessary properties of the aether were not like any known substance. Light was known to be transmitted as *transverse* waves, and not by the sort of longitudinal pressure waves that transmit sound. The term aether was just used to describe the propagation of electromagnetic waves. It was largely a figure of speech, like the term *lines of force* that was popular in the 1800s but which has now been replaced by *force field.*

If you think about light as being photon particles going through empty space, like bullets shot from guns, then your intuition might fool you. Bullets go faster if they are fired from a gun that is going forward. On the other hand, waves only go as fast as the medium allows. If a fire truck is driving towards you with the siren on, the sound of that siren does not get to you any faster. Sound is wave through air, and it only travels at a speed dependent on the properties of air, and not on whatever emitted the sound. You will hear the siren at a higher pitch, because of something called the *Doppler effect,* but you will not hear an approaching siren any sooner. You can think of the aether as being whatever medium it is that makes light waves all travel at the same speed.

Poincare seemed to be indifferent to the aether. He wrote in 1902 that he believed that the aether was perfectly undetectable, because attempts to detect it had failed and because any detection would violate his principle of relativity. He predicted that some day the aether would be discarded as useless. But he continued to talk about the aether as if it were a convenient hypothesis. In 1900, he denied that the aether could be detected by experiments like Michelson-Morley:

> Our aether, does it really exist? I do not believe that more precise observations could ever reveal anything more than relative displacements.

In 1889, and reprinted in his 1902 book, Poincare wrote that belief in the aether is just a convenience:

> Whether the ether exists or not matters little – let us leave that to the metaphysicians; what is essential for us is, that everything happens as if it existed, and that this hypothesis is found to be suitable for the explanation of phenomena. After all, have we any other reason for believing in the existence of material objects? That, too, is only a convenient hypothesis; only, it will never cease to be so, while some day, no doubt, the ether will be thrown aside as useless.

In other words, the aether is just a philosophical construct with no observable consequences. Lorentz expressed a similar opinion in 1914:

> The latter is, by the way, up to a certain degree a quarrel over words: it makes no great difference, whether once speaks of the vacuum or of the aether.[71]

Poincare lectured and published on electromagnetism and Maxwell's equations as early as 1888.[72] At the time, there were competing concepts regarding the role of the aether. Some argued, for example, that material objects could impart momentum into the aether, and that the aether would drift along with such objects. Poincare was persuaded by Lorentz's theory, and rejected arguments about the aether having some such mechanical function. He sometimes alluded to the aether theories of others, but his own explanations of Maxwell's equations were not dependent on an aether having any detectable function other than the propagation of electromagnetism.

Poincare's popular 1905 book[73] said, "to say the aether exists is to say there is a natural kinship between all the optical phenomena." Talking about the aether is just a way of talking about the transmission of light and other electromagnetic waves. Light and radio waves satisfy the same equations, and the aether is just a symbolic way of saying that they are just different frequency versions of the same thing.

Minkowski rarely mentioned the aether, as it was irrelevant to his geometric view. He noted that relativity was founded on the experimental failure to measure any motion relative to the aether, and he used the term in quotes when discussing Lorentz's treatment of Maxwell's equations. It was obvious that he was not using any preferred aether frame of reference because the whole point of his geometric approach was to relate every frame to every other frame with a symmetry group.

Maxwell was a proponent of the aether, as being necessarily to explain light and electromagnetic waves. When he first proposed that light was an electromagnetic wave, he brilliantly said that it was a wave in "the same medium which is the cause of electric and magnetic phenomena." He wrote the encyclopedia article on it in 1878, shortly before he died:

> The hypothesis of an aether has been maintained by different speculators for very different reasons. ... The only aether which has survived is that which was invented by Huygens to explain the propagation of light. The evidence for the existence of the luminiferous aether has accumulated as additional phenomena of light and other radiations have been discovered; and the properties of this medium, as deduced from the phenomena of light, have been found to be precisely those required to explain electromagnetic phenomena. ... We know that the aether transmits transverse vibrations to very great distances without sensible loss of energy by dissipation. ... Whatever difficulties we may have in forming a consistent idea of the constitution of the aether, there can be no doubt that the interplanetary and interstellar spaces are not empty, but are occupied by a material substance or body, which is certainly the largest, and probably the most uniform body of which we have any knowledge.[74]

Remember that the essence of Maxwell's theory is his system of equations. When he says that interstellar space is not empty, he means that it has the property that it transmits light in accordance with his equations. He was correctly saying that we do not know what the aether is, but we do know how it transmits light. And the uniformity of the aether turned out to be the essence of relativity.

Fresnel had an *aether drift* theory in 1818 that had some experimental success. The theory said that the aether could be pushed by the Earth, and flow like an ocean current. Lorentz rejects this in the introduction to his 1895 book, and says that he was avoiding speculations about the aether. He refers to the aether being at rest, but he explains that all he means by that was that he was rejecting the aether drift theory, and saying that it was meaningless to say whether or not the aether was really at absolute rest. After mentioning some hypotheses about how matter could pass through the aether, he says:

> It is not my intention to enter into such speculations more closely, or to express assumptions about the nature of the aether. I only wish to keep me as free as possible from preconceived opinions about that substance, and I won't, for example, attribute to it the properties of ordinary liquids and gases. If it is the case, that a representation of the phenomena would succeed best under the condition of absolute permeability, then one should admit of such an assumption for the time being, and leave it to the subsequent research, to give us a deeper understanding.
>
> That we cannot speak about an *absolute* rest of the aether, is self-evident; this expression would not even make sense. When I say for the sake of brevity, that the aether would be at rest, then this only means that one part of this medium does not move against the other one and that all perceptible motions are relative motions of the celestial bodies in relation to the aether.

In other words, Lorentz was avoiding the previous theories of aether motion, and the properties of the aether were extraneous to his analysis. The whole point of his 1892 paper was to deny that the aether could be used to detect the motion of the Earth. His 1904 paper also refers to the aether, such as saying that he deduced from Michelson-Morley "that the dimensions of solid bodies are slightly altered by their motion through the aether." But when he needs a frame of reference, he does not say that one is determined by the aether, and just says, "if we use a fixed system of coordinates".[75] Lorentz did not have a preferred frame in the sense of a frame with different physics. Sometimes it is argued that Lorentz and Einstein were working on entirely different theories, because Lorentz was preserving the aether while Einstein was abolishing it. But actually Lorentz rejected the previous aether theories as much as Einstein did.

FitzGerald was not bashful about the aether, and made colorful remarks such as this, from a 1900 lecture:

> We are harnessing the all-pervading ether to the chariot of human progress and using the thunderbolt of Jove to advance the material progress of mankind.[76]

Blaming FitzGerald for mentioning the aether is about as silly as blaming him for mentioning Jovian thunderbolts, or blaming modern astrophysicists for mentioning God.

The German physicist Paul Drude was the editor of the journal that published Einstein's famous papers, and he wrote a 1900 book on optics. He endorsed Lorentz's "elegant theory", and said, "the ether is conceived to be not a substance but merely space endowed with certain physical properties."[77]

Larmor wrote a 365-page book on *Aether and Matter* in 1900, and described the aether as "a mental construction or analogy, designed to relieve the mind from the intangible and elusive character of a complex of abstract relations."[78] The British mathematician Ebenezer Cunningham credited Larmor (1900), Lorentz (1904), and Einstein (1905) for the Lorentz transformations in 1907, and wrote:

> it is not permissible to speak of the velocity of an observer relative to the aether, as though the aether were a material medium *given in advance.* ... The aether is, in fact, not a medium with an objective reality, but a mental image which is only unique under certain limitations[79]

Einstein's famous 1905 paper used slightly different terminology. He said that attempts to detect the motion of the Earth relative to the "light medium" had failed, and that the aether was "superfluous" to his derivation. He then used the term "stationary system" as his convenient hypothesis. He used the "Maxwell-Hertz equations for empty space", while Lorentz discussed the same equations as the "equations for the aether". Einstein later made conflicting statements about the existence of the aether. In a 1909 review paper, he attempted to use the aether as a way of distinguishing his relativity from Lorentz's, and said, "Today, however, we regard the ether hypothesis as obsolete." He continued to deny the aether until he decided that it was needed for general relativity in 1916, and he affirmed it afterwards for the rest of his life. In a 1920 lecture on the aether, he said:

> More careful reflection teaches us, however, that the special theory of relativity does not compel us to deny aether. ... therefore, there exists an aether. According to the general theory of relativity space without aether is unthinkable; for in such space there not only would be no propagation of light, but also no possibility of existence for standards of space and time ...[80]

Einstein explained in a 1924 essay that Euclidean geometry does not need an aether, but special relativity requires an aether for spacetime, and the aether is absolute. General relativity also needs an aether, but that aether is not absolute.[81] In a 1919 letter to Lorentz, he adopted a view that was nearly identical to that of Lorentz's 1895 book:

> It would have been more correct if I had limited myself, in my earlier publications, to emphasizing only the non-existence of an aether velocity, instead of arguing the total non-existence of the aether, for I can see that with the word aether we say nothing else than that space has to be viewed as a carrier of physical qualities.[82]

Einstein explained Lorentz's belief in the aether this way:

> In view of his unqualified adherence to the atomic theory of matter, Lorentz felt unable to regard the latter as the seat of continuous electromagnetic fields. He thus conceived of these fields as being conditions of the aether, which was regarded as continuous. Lorentz considered the aether to be intrinsically independent of matter, both from a mechanical and a physical point of view. The aether did not take part in the motions of matter, and a reciprocity between aether and matter could be assumed only in so far as the latter was considered to be the carrier of attached electrical charges.[83]

In 1934 Einstein explained:

> Physical space and the aether are only different terms for the same thing: fields are physical states of space. If no particular state of motion can be ascribed to the aether, there do not seem to be any grounds for introducing it as an entity of a special sort alongside space.[84]

And he said later:

> We shall say: our space has the physical property of transmitting waves, and so omit the use of a word (ether) we have decided to avoid.[85]

He also seemed to deny the possibility of empty space at times. He sometimes explained general relativity this way:

> People before me believed that if all the matter in the universe were removed, only space and time would exist. My theory proves that space and time would disappear along with matter. There is then no 'empty' space, that is, there is no space without a field.[86]

As late as 1952, he said, "There is no such thing as an empty space". In a 1921 book on relativity, Pauli suggested defining the aether as "the totality of those physical quantities which are to be associated with matter-free space." He said that the aether exists with no space coordinates or velocities.[87] Einstein's view of the aether was not significantly different from the views of Lorentz in 1895 and the other relativity physicists before Einstein.

The modern aether

The aether has a colorful history that goes back to the ancient Greeks. Aristotle said that the aether was a fifth element, after earth, water, air, and fire. He said that, "Nature abhors a vacuum." The modern view is that the term "aether" has fallen out of favor, like the term *phlogiston*. Phlogiston was a word for the stuff that burns in a fire. We now have a much more sophisticated understanding of the chemistry of combustion and oxidation, and no one talks about phlogiston, but it is still correct that there is stuff that burns in fire, whatever you want to call it.

Likewise, there is an aether, whether you want to call it that or not. The famous physicist Paul Dirac, in a 1951 letter to Nature magazine, explained that quantum electrodynamics requires an aether.[88] It just uses different terminology. Instead of waves in the aether, the modern theory describes perturbations in the electromagnetic quantum vacuum. Other terms for electromagnetic properties of what seems like empty space are the *Dirac sea, impedance of free space, magnetic permeability of free space, zero-point energy, vacuum energy, quantum foam,* and *Casimir effect*. The term aether could also be used for *dark energy, quintessence, chiral condensate,* or *CMB radiation*, as explained below. You could say that 20th century physics requires an aether, although it is a little different from what people expected in 1900. Our concepts of atoms, light, gravity, and a lot of other things are different from what people had in 1900, so there is nothing wrong with continuing to use the term aether. As Whittaker's 1953 textbook explains:

> As everyone knows, the aether played a great part in the physics of the nineteenth century; but in the first decade of the twentieth, chiefly as a result of the failure of attempts to observe the earth's motion relative to the aether, and the acceptance of the principle that such attempts must always fail, the word 'aether' fell out of favour, and it became customary to refer to the interplanetary spaces as 'vacuous'; the vacuum being conceived as mere emptiness, having no properties except that of propagating electromagnetic waves. But with the development of quantum electro-dynamics, the vacuum has come to be regarded as the seat of the 'zero-point' oscillations of the electromagnetic field, of the 'zero-point' fluctuations of electric charge and current, and of a 'polarisation' corresponding to a dielectric constant different from unity. It seems absurd to retain the name 'vacuum' for an entity so rich in physical properties, and the historical word 'aether' may fitly be retained.[89]

This statement is even more true today, as additional physical properties of the vacuum/aether have been discovered. It is a consequence of relativistic quantum field theories that virtual particles are always getting created and annihilated throughout a vacuum. These virtual particles give properties to the vacuum that you would not expect from empty space. Electrons appear to be less charged because of *vacuum polarization*. The best modern theory of light and electrons, quantum electrodynamics, can be regarded as a perturbation theory for the aether.

A basic premise of quantum mechanics is that electrons and other particles have wave-like properties that prevent them from being isolated with zero momentum at a particular position. This is called the *Heisenberg uncertainty principle,* after the German physicist Werner Heisenberg. The principle also implies that zero energy can never be achieved. Even a vacuum must have energy fluctuations.

The theories of the strong and weak nuclear forces similarly require a complicated vacuum. So in a sense, relativity requires an aether because it requires the vacuum to have material properties different from empty space. The aether is relativistic and looks the same in each frame, so it is consistent with rejecting aether velocity instead of the aether, as Einstein wished that he had advocated. As Maxwell said, the aether is "the most uniform body of which we have any knowledge."

Quantum electrodynamics teaches that there is no such thing as empty space, and there is no known way for light to propagate through empty space even if there were such a thing. The theory uses the word *vacuum* to mean the lowest energy state of the system, not for empty space. Using the word *aether* for the quantum vacuum is consistent with the aether that Maxwell, Larmor, and Lorentz wrote about in the late 1800s. Either way, light or any other electromagnetic wave is best understood as a perturbation in an esoteric structure that uniformly and invisibly permeates all of space and time. While that structure is not always called the aether, it is one of the most universally accepted concepts in all of science.

The modern aether theory is not just some philosophical construct. It is an essential part of the most modern, fundamental, accurate, and widely accepted theory in all of physics. As the MIT theoretical physicist and Nobel prizewinner Frank Wilczek explains:

> Quite undeservedly, the ether has acquired a bad name. There is a myth, repeated in many popular presentations and textbooks, that Albert Einstein swept it into the dustbin of history. ... the truth is more nearly the opposite ... At present, renamed and thinly disguised, it dominates the accepted laws of physics.[90]

Wilczek also says that the idea that Einstein eliminated the aether is a "vulgar misunderstanding".[91] He says that "the concept that what we ordinarily perceive as empty space is in fact a complicated medium", and that it "is in reality a wildly dynamical medium".[92] He also calls it the "symmetry-breaking aether",[93] and says that it is "the primary ingredient of physical reality" and the "primary reality" for modern physics.[94] While the aether has a symmetry under the Lorentz group, the mirror reflection symmetry is broken so that the aether is slightly more right-handed than left-handed. Some other symmetries are also broken. How those symmetries are broken or unbroken underlies much of our understanding of modern particle physics.

Without the aether, there is no good explanation for why all electrons should appear identical. In ordinary non-relativistic mechanics, it seems like an unlikely coincidence for particles to be similar. Wilczek says:

> Undoubtedly the single most profound fact about Nature that quantum field theory uniquely explains is *the existence of different, yet indistinguishable, copies of elementary particles*. Two electrons anywhere in the Universe, whatever their origin or history, are observed to have exactly the same properties. We understand this as a consequence of the fact that both are excitations of the same underlying ur-stuff, the electron field. The electron field is thus the primary reality. The same logic, of course, applies to photons or quarks, or even to composite objects such as atomic nuclei, atoms, or molecules. The indistinguishability of particles is so familiar, and so fundamental to all of modern physical science, that we could easily take it for granted. Yet it is by no means obvious.[95]

The "ur-stuff" is yet another euphemism for the aether. His later book calls it the "grid". The electron is the quantization of a uniform Poincare-symmetric field that pervades all of space and time. The electrons are all the same because they are just the smallest observable units of an aether that is symmetric in space, time, and motion. The aether is the largest and most uniform body in the universe, just as Maxwell concluded in the 1878 encyclopedia.

The uniformity of the aether is another way of stating Einstein's postulates. If the aether is the medium for light and has no detectable direction or velocity, then the speed of light must be the same for all observers. From this the Lorentz transformations may be deduced, as Lorentz did in 1892-1904 and Einstein did in 1905. The uniformity of the aether is also another way of stating the outcome of the Michelson-Morley experiment. Their apparatus always showed the same light patterns, no matter what the direction or velocity.

Modern physics depends on the aether as the medium for electromagnetism. You might say "the medium is the message", as the aether medium is the most fundamental part of modern field theory. That slogan was introduced by a 1964 Marshall McLuhan book,[96] and was meant to emphasize the importance of the mass communications media like radio and television.

One can take the view that electromagnetic waves do not need a medium, and that the aether is synonymous with just empty space (or empty spacetime). But that view is inadequate to explain the speed of light being constant, or all electrons having the same mass and charge. Electromagnetism has properties that are uniform and everywhere, and they can either be described in terms of a pervasive aether, or in terms of physical laws that are functionally equivalent to such an aether.

The *Casimir effect* is perhaps the most directly observable evidence that the vacuum is not really empty space. It was predicted by the Dutch physicist Hendrik Casimir in 1948, and measured in subsequent decades. Suppose you place two metal plates only a few microns apart in a vacuum. There are no charges or magnets, so there should be no electromagnetic fields to cause forces. You can cool it down to a temperature near absolute zero so that there will be no thermal effects, and make the plates light enough that there will be no measurable gravitational effects. And yet there is a measurable force between the plates, and it is explained in terms of the quantum electrodynamics of the vacuum. It is possible that this force is the same is the same as the dark energy that is accelerating the expansion of the universe.

The whole modern theory of particle physics is based on the vacuum having a universal aether-like invisible field that gives particles mass. That field has not been observed directly, but it is very widely accepted for reasons explained later. There is no other good theory about why particles can have masses in a relativistic world. There is also a universal dark energy that is widely accepted as causing the expansion of the universe to accelerate. It may or may not be the same as one of these other aether concepts.

The most common cosmological models today do have a universal rest frame, and the rest frame is accepted without controversy. The rest frame is the frame of the Cosmic Microwave Background (CMB), and is sometimes call the *comoving* frame. The measure of time in this special frame is called *cosmological time*. It is the frame that allows us to say that the universe is nearly 14 billion years old. Special satellites have collected data on radiation leftover from the *big bang* that created the universe, and radiation moving towards us appears slightly warmer. From that data, our Milky Way galaxy appears to be moving at a speed of 627 ± 22 km/sec relative to the CMB rest frame. The 2006 Nobel Prize was given for this work, and the acceptance lecture called it the "new aether drift experiment".[97] Therefore it is not correct to say that there can be no aether because there can be no preferred rest frame, because there really is a cosmologically-defined notion of being at rest. Astronomers no longer just talk about relative motion of stars, and use the CMB to measure the *systemic motion*. The motion relative to the nearby stars is called the *particular motion*. If we received a message from an alien civilization on another planet, we could understand a statement about systemic motion because they could refer to the same universal rest frame. We could also understand a statement about time, as they could measure the age of the universe in cosmological time just the same as we do.

The curious issue is why anyone would attach such great importance to Einstein's opinion about the aether, if it had no known observational consequences anyway.

Einstein's second most famous paper, also published in 1905, is credited with discovering that light is composed of particles, now called photons, instead of waves. The German physicist Max Planck had already proposed, in 1905, that light was absorbed and emitted in discrete quanta, with energy proportional to frequency. Red light has a frequency of about 480 terahertz, and blue light is about 650 terahertz, so a blue photon has about 35% more energy than a red photon. You get the energy multiplying by what is now called *Planck's constant*. Einstein gave a "heuristic" argument that light should be considered as being transmitted through space as discrete quanta as well.

The modern view is that light looks like a particle when it is observed, that is when it is absorbed or emitted, and it otherwise behaves like a wave. If you ask whether light is really a wave or a particle, you will not get a good answer. Some will say that it is a particle that sometimes acts like a wave, and some will say that it is wave that sometimes acts like a particle. Some will say that it is both, and some will say that it is neither. Some will give you a 500-page book on quantum field theory, and some will say that it is one of those quantum mysteries that will never be understood.

The same is true about electrons and other fundamental particles. Whenever an electron is observed, it looks a point particle of zero diameter that is identical to any other electron. But when you are not looking, an electron satisfies a wave equation. That wave equation can be solved to predict that the electrons in a molecule will be shaped into orbitals, and then the shape of those orbitals can be used to deduce the chemical properties of that molecule. Much of chemistry depends on understanding electrons as waves, not particles, and yet electrons are always observed as particles.

Modern physics is filled with paradoxes such as this. Very often there is more than one way to understand some physical phenomenon, even though those understandings may have very different premises. When that happens, it is often impossible to say whether those premises are correct in the sense of truly representing reality. Perhaps you can only say that the premises are useful hypotheses.

Quantum mechanics is especially confusing with its different interpretations. Everyone agrees that the theory is superb at predicting experiments, with observation agreeing with theory to as many as 8 or 9 decimal places. But when you ask what it all means, you get wildly different answers.

As before, there is no agreement on which is more fundamental, the particles or the fields. Beyond that, there is the curious nature of the *Copenhagen interpretation* of quantum mechanics, developed in the 1920s by Danish physicist Niels Bohr and others. Under this interpretation, the Moon might only exist when you look at it, and Schroedinger's cat might be half dead and half healthy in a box. Bohr said that the particle and wave models were complementary, so electron and photons sometimes look like particles, and sometimes like waves. He also said that truth and clarity are complementary.

The Copenhagen interpretation is so goofy that many physicists will deny that they believe in it, until you ask them what they do believe instead. Another interpretation is the *many-worlds theory*, or *multiverse*. In it, every possible scenario, including the most farfetched movie plots, are actually being played out in some other universe. Another one is the *alternative histories* interpretation, where events have many histories and certain possibilities become inaccessible somehow.

So far, these interpretations have more or less all the same observational consequences. That is why they are called interpretations, and not theories. You can pick and choose whichever you care to believe, and no one can prove you wrong.

There is a lot of active research on trying to understand these interpretations better. Under some interpretations of quantum mechanics, it would be possible to build a quantum computer that would be vastly more powerful than existing computers on certain kinds of problems that can be solved by doing many computations in parallel.

If such a computer gets built, then we will gain a better understanding of quantum mechanics. But so far, the researchers only claim that they have factored the number 15, and have not clarified the basic problems.

You might think that scientists should keep an open mind when different feasible explanations are available, but the lesson that people have learned from Einstein is that there is great glory in staking out a position on some untestable hypothesis, such as the non-existence of the aether.

Relativistic mass

Lorentz published a startling prediction in 1899. He said that the mass of a particle increases with velocity, as a consequence of his relativistic electrodynamics. It was another bold attack on classical mechanics.

Up to that point, mass was not known to ever change. Mass seems to disappear when you burn wood, but the French chemist Antoine Lavoisier had figured out a century earlier that it literally goes up in smoke. He made many brilliant chemical discoveries, but was guillotined during the French Revolution. The mass is conserved if you measure all of the combustion byproducts, including the smoke.

Mass has been defined since Newton in terms of inertial resistance to acceleration. The more massive an object is, the greater force is needed to accelerate it. So if mass increases with velocity, then a rapidly-moving particle becomes unusually resistant to acceleration. This had actually been observed by some physicists before Lorentz, but no one could explain it. Lorentz gave formulas for the increase, as a consequence of relativity. He perfected his derivation in 1904, and gave formulas that work for all velocities less than the speed of light. Einstein gave similar formulas in 1905.

In relativity, nothing goes faster than the speed of light. So if a particle gets close to the speed of light, then it must become more resistant to acceleration, because even very large forces will not make the particle go faster than light. That is how Lorentz's relativistic mass worked. It gave a dynamical explanation as to why you can never get faster than light.

Today it might seem obvious that electric charge would be carried by particles and that those particles must have mass, but it was not so obvious in the 1800s. While physicists believed in atoms, no one had any way of saying anything about the charge or mass of an electron, if even such a thing existed. Lorentz was the first to give such a theory.

Lorentz's student Peter Zeeman did a remarkable experiment in 1896 that showed that a strong magnetic field could cause a splitting in the colors of emitted light. He was fired for doing the work as no one but Lorentz appreciated its significance. Lorentz explained it in terms of his electron theory, and deduced that electrons have mass equal to about one thousandth of the mass of a hydrogen atom. Zeeman got a Nobel Prize in 1902 for this work, along with Lorentz.

Lorentz's mass was strange in that it was no longer a simple number. It was harder to accelerate a particle in the direction of motion than the transverse direction, so he had a concept of *longitudinal* mass and *transverse* mass. He introduced these terms in his 1904 paper:

> These quantities m1 and m2 may therefore properly be called the "longitudinal" and "transverse" electromagnetic masses of the electron.

Lorentz's electron theory was the basis for the 1897 invention of the cathode ray tube. Such tubes were used on all the non-flat-screen television sets and computer monitors until they became obsolete several years ago. They use a magnetic field to control the way an electron beam hits a fluorescent screen. The Lorentz force law described how much a magnetic field would deflect an electron beam, and he said that relativistic mass would make the electrons more resistant to deflection. The German physicist Walter Kaufmann and others did experiments that convinced everyone that the electron beam consists of particles, and FitzGerald proposed that they be called electrons.

Relativistic mass was tested in a series of experiments starting in about 1901. The results were qualitatively consistent with Lorentz's formulas, but not sufficiently accurate to distinguish other theories.

Kevin Brown argues that Einstein's 1905 view had physical advantages over Lorentz's theory:

> To give just one example, we may note that prior to the advent of special relativity the experimental results of Kaufmann and others involving the variation of an electron's mass with velocity were thought to imply that all of the electron's mass must be electromagnetic in origin, whereas Einstein's kinematics re-

> vealed that all mass - regardless of its origin - would necessarily be affected by velocity in the same way.[98]

There is a popular misconception that Einstein was the first to predict changes in length, time, and mass at high velocities and that those changes will only be tested when we build Star Trek spaceships. In fact the predictions about relativistic mass were already being tested in 1901, before Einstein wrote anything on the subject. Kaufmann did experiments showing that electron mass increases with velocity.

Kaufmann was not claiming to have confirmed the Lorentz theory. He supported a rival theory that electrons had some sort of electromagnetic mass differing from regular mass. In retrospect, his experiment was precise enough to show a mass increase with velocity, but that is about all. It was a qualitative confirmation of relativity. It was not until around 1908 that experiments were accurate enough to show that Lorentz's formula was better than the alternatives.

So there was a debate in 1905 about whether the relativistic mass of an electron was electromagnetic in origin. Let's look at what Einstein says in his famous paper. He discusses mass in section 10, which starts:

> Let there be in motion in an electromagnetic field an electrically charged particle (in the sequel called an "electron"), for the law of motion of which we assume as follows:--

So he is talking about electromagnetism. He does not do the derivation in sections 1-5, which are devoted to kinematics, independent of electromagnetism. He does argue that the mass effect applies more broadly:

> With a different definition of force and acceleration we should naturally obtain other values for the masses. This shows us that in comparing different theories of the motion of the electron we must proceed very cautiously.
>
> We remark that these results as to the mass are also valid for ponderable material points, because a ponderable material point can be made into an electron (in our sense of the word) by the addition of an electric charge, no matter how small.

But he continues to talk about electromagnetism only for the rest of the section. So Einstein's approach is quite similar to Lorentz's earlier papers — using electromagnetism to derive the mass effect, and then claiming that the principle applies more broadly. Einstein used terminology that seemed to be quoting Lorentz's 1904 paper, even though he always claimed that he had not seen that paper:

> Taking the ordinary point of view we now inquire as to the "longitudinal" and the "transverse" mass of the moving electron.

Today, the term *relativistic mass* has fallen out of favor. A moving object does appear to have an increase in mass, in the sense that there is an increase inertial resistance to forces. However the modern relativity textbooks say that this terminology is confusing, because the inertial resistance is greater in the direction of motion than perpendicular to the direction of motion. Lorentz (in 1899 and 1904) and Einstein (in 1905) dealt with this difficulty by using the concepts of longitudinal mass and transverse mass. But that would mean that mass has a direction associated with it, whereas we think of mass as just being a single real number. So most physicists prefer to avoid relativistic mass and discuss the effect in terms of *momentum*, as momentum is a 4-component 4-vector and mass is a just a single number.

In spite of this terminology problem, the relativistic mass effect is real, and it applies to any particle or object, whether electromagnetism is involved or not. The idea that there are two kinds of mass, mechanical and electromagnetic, has been firmly rejected. The relativistic mass effect is a consequence of momentum being a 4-component vector on a 4-dimensional spacetime with a Poincare symmetry group.

It was Poincare, not Einstein, who firmly rejected the wrong idea that electromagnetic mass responds to motion differently from mechanical mass. Before Einstein, Poincare had already addressed the issue directly and correctly in his 1904 St. Louis speech the previous year:

> Now, the calculations of Abraham and the experiments of Kaufmann have shown that the mechanical mass properly so called is nothing, and that the mass of the electrons, at least of the negative electrons, is purely of electrodynamic origin. This is what compels us to change our definition of mass; we can no longer distinguish between the mechanical mass and the electrodynamic mass, ... it is necessary, therefore – so [Lorentz] says – that the masses of all particles be influenced by a translation in the same degree as the electromagnetic masses of the electrons. Hence, the mechanical masses must vary according to

> the same laws as the electrodynamic; they can then not be constant.

Poincare more boldly and correctly attacked conventional wisdom on this matter, and he did it a year ahead of Einstein.

Einstein's defenders often attack Poincare for using terms like "apparent mass". They say that this shows that he did not understand that the physical effects were real, or else he would not have used terms like "apparent". But Poincare's terminology is very similar to today's relativity textbooks that are reluctant to say that the mass increases with velocity. Yes, the mass appears to increase but there are pedagogic reasons for not calling it a mass increase. The term "apparent mass" is a perfectly good term.

Mass energy equivalence

The most dramatic consequence of relativity is the atomic bomb. Einstein discovered the most famous equation ever found, $E=mc^2$, and that was the theory behind the atomic and hydrogen bombs. Or so the story goes. Actually, Einstein did not invent this formula or the atomic bomb. The formula is not even needed for the atomic bomb.

The big idea is that mass and energy are really the same thing. Mass can be converted to energy and energy can be converted to mass. In the formula, E is energy, m is mass, and c is the speed of light. The speed of light is a big number, so a little bit of mass can generate a huge amount of energy, as in an atomic bomb. Conversely, converting energy to mass will result in so little mass that you are unlikely to notice.

In classical mechanics, an object of mass m and velocity v has momentum mv and kinetic energy $.5mv^2$. If you apply a force to accelerate an object, the work you do increases the object's momentum and energy according to these formulas. Relativity teaches that nothing goes faster than light, so it appears that no matter how much energy you put into accelerating an object, its momentum is always less than mc and its kinetic energy is always less than $.5mc^2$. But that is not correct. There are no such limits.

Momentum and energy are conserved quantities, whether using classical or relativistic mechanics. The energy that is put into that object must go somewhere. Either the formulas for momentum and kinetic energy must be modified for relativity, or the mass increases with velocity. In other words, the energy gets turned into mass.

One of the most fundamental principles of physics is the conservation of energy. If we could create energy out of nothing, then we could make perpetual motion machines and solve many of the world's problems. If we could do an experiment that destroyed energy, then maybe we could run the experiment backwards to create energy. The history of science tells us that the creation and destruction of energy is impossible.

Preserving the principle of energy conservation has required inventing new forms of energy. After all, every time you strike a match, you seem to create heat and light energy where there was none before. So that there is no contradiction, the notion of chemical potential energy was invented. The flame's energy is explained by saying that the energy is locked up in the chemicals in the match, and then released when you light the match. Likewise, when you drop a rock, you are releasing gravitational potential energy. Physics has a history of recognizing different kinds of energy so that total energy is conserved.

Likewise, relativity requires that mass is another form of energy, so that energy is conserved. The c-squared in the formula is just a distraction. Relativity teaches that $c = 1$ in convenient units. That is what Poincare used in his long 1905 paper. Saying $c = 1$ is an expression of relativity, because it says that the speed of light is constant and measuring space and time are related. So the famous formula is really just the statement that energy equals mass, in appropriate units, and that mass is just another form of potential energy.

When a uranium atom is split into two smaller atoms, the sum of the masses of those smaller atoms does not quite add up to the mass of the original uranium atom. The difference gets turned into energy according to $E=mc^2$. That is the only relationship between relativity and the atomic bomb. Relativity does not give any clue on how to split an atom, or how to create a nuclear chain reaction, or any of the other necessary steps to making an atom bomb. Relativity is not even needed to understand the energy release in a uranium or plutonium bomb, as the release can be largely explained from electromagnetic considerations.

The electromagnetic explanation for fission energy is quite simple. If a uranium nucleus (92 protons) splits into an yttrium nucleus (39 protons) and an iodine nucleus (53 protons), then the energy released is equal to the electrostatic energy required to push those positively charged nuclei close together. Like charges repel, and the closer they are together the more they repel, according to an inverse square law. Calculating the fissionable energy of uranium is like calculating the potential energy of a compressed spring; if you know the force of a spring then you can compute the work required to compress the spring.

The atomic bomb was developed by the American Manhattan project during World War II. Einstein was not asked to participate, as his skills were obsolete by then. His role was only to sign the letter to the President. He would not have gotten the necessary security clearances anyway, as he was having an affair with a Soviet spy at the time.[99]

Nevertheless, Einstein and the formula are popularly identified with the atom bomb. The bomb became public when it was used on Japan at the end of World War II. The next day the New York Times reported on the front page that it "was the first time that Prof. Albert Einstein's theory of relativity has been put to practical use outside the laboratory."[100] The next year, Einstein, the formula, and a mushroom cloud were all together on the cover of Time magazine.[101]

The formula $E=mc^2$ was needed to explain the energy released in hydrogen (fusion) bomb. The energy for that comes from the strong (nuclear) force, which was not understood when the first bombs were built. A true relativity bomb would be to combine matter with antimatter, thereby converting *all* of the mass to energy. It is possible to use a positron to annihilate an electron in a lab experiment, releasing pure energy in the form of radiation (i.e., photons). Such annihilations occur in medical imaging equipment called *positron emission tomography* (PET) scanners. But there is no way (outside of a fictional Dan Brown novel) to bottle enough antimatter for a bomb.

After Einstein's famous 1905 paper on special relativity, he wrote a short follow-up paper with his famous formula. He expressed it as:

> If a body gives off the energy E in the form of radiation, its mass diminishes by E/c^2.

This paper also had no references and was not as original as it appeared. Lorentz had already proposed relativistic mass in 1899, and experiments had begun to observe it in 1901. In a 1900 paper Poincare wrote:

> Therefore, from our point of view, since the electromagnetic energy behaves as a fluid which has inertia, we must conclude that, if any sort of device produces electromagnetic energy and radiates it in a particular direction, that device must recoil just as a cannon does when it fires a projectile.

He then gives a numerical example of the mass being changed by E/c^2. Poincare also made use of this mass-energy equivalence in his 1905 papers.

The Austro-Hungarian physicist Friedrich Hasenohrl published a paper in 1904 giving an argument that mass due radiation is $(8/3)E/c^2$. He partially fixed his mistake and said half that value in a follow-up paper in the very same journal where Einstein later published his famous papers. He said that it was consistent with the contraction of Lorentz and FitzGerald.

Einstein spent his whole life denying that he was influenced by Poincare, and hardly ever even acknowledged him. But in a rare example, an Einstein 1906 paper credited Poincare's 1900 paper for $E=mc^2$. He wrote:

> In the present paper I want to show that the above theorem [$E=mc^2$] is the necessary and sufficient condition for the law of the conservation of motion of the center of gravity to be valid (at least in first approximation) also for systems in which not only mechanical, but also electromagnetic processes take place. Although the simple formal considerations that have to be carried out to prove this statement are in the main already contained in a work by H. Poincare, [the previous mentioned paper] for the sake of clarity I shall not base myself upon that work (5, p. 252)

It appears that Einstein was concerned about who might get the credit. He never credited Poincare again,[102] and he never credited Hasenohrl's papers. He never mentioned Olinto De Pretto who published the formula in a 1903 science magazine (but without a relativistic derivation). He resented the possibility that Planck might get credit, as he complained to Stark in 1908:

> I was rather disturbed that you do not acknowledge my priority with regard to the connection between inertial mass and energy.

Some people argue that Einstein's innovation was to give a proof of the formula, but that is not really correct. While it is possible to extract a proof from the idea in Einstein's paper, he used low-velocity approximations and circular logic. In the succeeding years, others wrote papers with more convincing arguments. Einstein went on to publish seven different proofs of $E=mc^2$ over the course of forty years, but all of them were incorrect.[103]

Since the concept of relativistic mass has fallen out of favor, many physicists prefer to equate mass with rest mass. For them, $E=mc^2$ is only true for an object at rest, and the correct formulation is that the energy E of an object with rest mass m and momentum p is given by $E^2=(mc^2)^2+(pc)^2$. Planck published this formula in 1906, and showed that it could be used as a basis for mechanics.[104] Momentum is not just mass times velocity, but relativistic momentum. Planck's formula is a generalization of the rest mass case because $p=0$ when the object is at rest. The formula also has the special case that $E=pc$ for a particle with zero rest mass, like the photon.

Planck's formula looks like the Pythagorean theorem in disguise, and it is. A vector on spacetime has four components, three spatial and one time. Energy-momentum is really just one vector with four components. If a particle is moving with momentum p in the direction of the x-axis, then its energy-momentum vector has components $(p,0,0,E/c^2)$. If it were at rest (in that coordinate system) then the vector would be $(0,0,0,m)$. This assumes that $E=mc^2$ in the rest frame. These vectors must have the same value of the metric, $0^2+0^2+0^2-c^2m^2 = p^2+0^2+0^2-c^2(E/c^2)^2$. Rearranging this gives the above formula.

Spacetime dynamics

Newton's great achievement was a theory of dynamics. He explained how gravitational and other forces affect motion. His famous second law said that force equals the rate of change of momentum. A force applied to an object over some distance does *work* that imparts energy. Force is what causes mass to accelerate. These ideas formed the core of the theory that allowed him to predict the motions of everything from the planets to everyday objects here on Earth.

Force and momentum are vectors. Newton did not know about vectors, but by 1900 it was understood that vector analysis provided the best formalism for classical dynamics. Every force has a particular direction in 3-dimensional space, and a magnitude. Likewise, momentum has a direction and magnitude.

It turns out that momentum and energy can be combined to make a 4-vector, and that a version of Newton's second law holds for the energy-momentum 4-vector. That is, the force 4-vector is equal to the rate of change of the energy-momentum 4-vector. The actual dynamics under relativity are a little different, because nothing ever goes faster than the speed of light, no matter how much force is applied.

Poincare figured out in 1905 how to extend forces to be 4-vectors, and how to convert Newtonian dynamics to give a relativistic prediction of planetary orbits.[105] He used these ideas to show that orbits were still approximately elliptical under relativity. He later used them to argue that relativity may explain why Mercury's orbit deviates from an ellipse. Minkowski and others further developed the spacetime dynamics.

Poincare also gave a relativistic version of the Lorentz force law, which gives the dynamics of an electron in an electromagnetic field. Lorentz had done something similar in 1904.

Einstein emphasized the kinematics in his 1905 paper, so he ignored Newtonian forces. He did give a treatment of the Lorentz force law, but only in a low velocity approximation. He considered forces in 1907, and gave a clever approximate argument about how acceleration would affect clocks. He later contributed to general relativity, where spacetime itself is dynamic.

Logical foundations of relativity

While Einstein scholars concede that the special relativity formulations of Lorentz and Poincare were mathematically equivalent to Einstein's, many argue that the beauty and depth of Einstein's version rests on its axiomatic foundations. Einstein showed in 1905 how the whole theory follows from two simple postulates. He assumed that motion is relative and the speed of light is constant, and then showed that all of the strangeness of relativity was a logical consequence. It is not really so simple.

The word *postulate* comes to us from Euclid of Alexandria, a Greek mathematician who lived around 300 BC. His book, *Euclid's Elements,* put geometry on solid logical foundations, and was the most influential book in the history of mathematics. It created a standard for mathematical proof.

Euclid understood that arguments have to depend on premises, and in mathematics, proofs of theorems have to depend on postulates. He had to assume, for example, that two points determine a line. His *Fifth Postulate* said that given a point and a line on a plane, there is exactly one parallel line through the point. Not until the 1800s did mathematicians understand that there were non-Euclidean geometries that satisfied all of the postulates except the fifth.

The great achievement of Euclid's Elements was that he was able prove startling theorems like the Pythagorean theorem from his postulates. These theorems became indisputable mathematical truths, because they had to be true unless there was some fault in either the postulates or the proofs.

While Einstein's postulates appear to have the self-evident simplicity of Euclidean postulates, they have a much more complex meaning. The first clue is in Einstein's own words. When he first defined his principle of relativity as a postulate in his famous 1905 paper, he said, "as has already been shown to the first order". He gave no reference to what had been shown, but in later interviews, he explained that he was using Lorentz's 1895 paper. Einstein meant that he was taking what Lorentz proved, and adopting it as a postulate.

Lorentz proved his theorem of the corresponding states for first order in velocity in 1895, and extended it to all velocities in 1904. Einstein always claimed in interviews that he did not know about Lorentz's 1904 paper, and that he independently conjectured that Lorentz's theorem was true for all velocities. So Einstein's first postulate was essentially the same as Lorentz's theorem.

Lorentz had to make his own assumptions, of course. His theorem was that there was correspondence between electromagnetic states in different frames of reference. The proof was given by applying his famous transformations to Maxwell's equations. He also explained the physical significance in terms of experiments about the motion of the Earth.

Minkowski also wrote about Lorentz's theorem and postulate.[106] To Minkowski, the theorem was a mathematical statement about the transformations of Maxwell's equations, and the postulate was the hypothesis that the formulas were physically realized. He credited Einstein with having "brought out the point very clearly" that Lorentz's postulate was not artificial, and with having "succeeded to some extent in presenting the nature of the transformation from the physical standpoint." Minkowski proved Lorentz's theorem, as well as covariance.

Einstein's second postulate was that the speed of light is a constant, independent of the motion of the source. Maxwell had proved it from his theory back in 1865. So both of Einstein's postulates can be proved from Maxwell's equations and the physical interpretation of the Lorentz transformations.

Einstein does avoid Maxwell's equations to derive his kinematics in the first half of his paper. But there are other hidden assumptions. For example, he says, "Let us take a system of co-ordinates in which the equations of Newtonian mechanics hold good." Relativistic kinematics are contrary to Newtonian mechanics, so there are no such coordinates. In the second half, Einstein has to assume a correspondence of electromagnetic variables, just as Lorentz had proved.

Lorentz compared his relativity to Einstein's by saying, "the chief difference being that Einstein simply postulates what we have deduced".[107] Modern textbooks do not describe Einstein's contributions this way, but that is how Lorentz, Einstein, and Minkowski described them at the time.

Poincare and Minkowski proved even better versions of Lorentz's theorem, and those later became known as Lorentz covariance. While everyone credits Einstein with achieving a certain simplicity with his postulates, Lorentz, Poincare, and Minkowski were doing something more ambitious. In short, they were proving what Einstein was postulating.

Poincare gave other derivations of relativity as well. He derived the Lorentz transformations from a least action principle, from his metric tensor, and from imaginary time rotations. He gave proofs as well, and wrote that "the postulate of relativity may be established with perfect rigor." Minkowski refined his ideas to present relativity in terms of 4-dimensional geometry. Thereafter, the preferred presentation of relativity has been in terms of postulates for 4-dimensional non-Euclidean geometry, as they avoid the Maxwell theory that was essential to Lorentz and Einstein. Electromagnetism is then explained in terms of Poincare covariance.

Motion of the Earth

By 1900, the motion of the Earth was well established. Every schoolchild knew that the Earth rotated daily, and revolved around the Sun annually. But some optical experiments (including Michelson-Morley) in the 1880s failed to detect any Earth movement, and relativity had to be developed to explain the anomaly.

Most people think that the motion of the Earth was proven by the German-Polish astronomer Nicolaus Copernicus in his famous 1543 book or by the Italian physicist Galileo Galilei in his 1633 confrontation with the Pope. A (doubtful) legend says that Galileo defiantly declared "and yet it moves" after the Inquisition forced him to recant his theory. They find it surprising that as late as 1887, an experiment could indicate that the Earth was stationary, and nobody had a good explanation for it.

Lorentz developed his relativistic transformations directly to explain the Michelson-Morley experiment. Poincare was led to his Principle of Relativity from Michelson-Morley also. They said so in their papers. They were nearly alone in understanding that the experiment could not be reconciled with the prevailing interpretation of Maxwell's equations, and that a new theory of space and time was needed. Most physicists believed that either the aether was being carried along with the Earth, or that electromagnetic and optical experiments would be able to detect the motion of the Earth through the aether.

Poincare explained some of his ideas in a popular 1902 book. His chapter on "relative and absolute motion" denied that there is any absolute space to give the motion of the Earth some objective existence. It said:

> Just as our Copernicus said to us: "It is more convenient to suppose that the earth turns round, because the laws of astronomy are thus expressed in a more simple language," ... these two propositions, "The Earth turns round," and "It is more convenient to suppose the Earth turns round," have the same meaning. There is nothing more in the one than in the other.[108]

The book caused quite a stir in Europe. It seemed to be saying that the Catholic Church was right in its dispute with Galileo. If he were not a famous mathematician, he might have been dismissed as a crackpot. He was boldly describing relativity theory to the general public, without equations. He was not trying to revive a centuries-old debate, but to make the point that motion is relative. There was no known objective way to say whether or not the Earth was in motion.

Poincare argued that if the Earth had a permanent thick cloud cover, so that no one even knew about the stars, then it would have taken a whole lot longer for someone like Copernicus to suggest the motion of the Earth. Scientists would have assumed a motionless Earth, and would have invented fictional forces to explain the flattening at the poles, Foucault's pendulum, and the rotational direction of cyclones. Eventually some clever physicist would declare that the laws of mechanics could be described more conveniently if the Earth were assumed to rotate, but he would be unable to prove that the Earth really does rotate.

Maxwell realized that there was a tension between the motion of the Earth and his theory of electromagnetic fields. It was easy to understand how material objects like rocks, water, and air would be carried along with the Earth, but not so clear whether the motion should have an affect on invisible fields. His theory opened up the possibility of an optical or electromagnetic experiment to detect the motion of the Earth relative to the aether. He did his own optical experiment in 1868, and failed to detect any motion. He even did it at different times of the year, because the Earth would be going in a different direction. His 1878 encyclopedia article had a whole section on experimental attempts to measure the relative motion of the aether. Michelson got the idea from him, and did better experiments.

Since Michelson-Morley was so crucial to relativity, it had long been assumed that it was a major motivator for Einstein. But Einstein just relied on Lorentz's and Poincare's deductions, and did not need the experiment. It is not even clear that he even knew about Michelson-Morley. He later said, "The Michelson-Morley experiment had no role in the foundation of the theory. ... the theory of relativity was not founded to explain its outcome at all." His famous 1905 paper did refer to "unsuccessful attempts to discover any motion of the earth", but did not mention Michelson. Einstein later made conflicting statements on the matter, and historians say that he did not rely on Michelson-Morley.[109] The obvious explanation is that Lorentz and Poincare relied on the Michelson-Morley experiment, and Einstein relied on them.

Einstein's 1905 paper does suggest an experiment to detect the rotational motion of the Earth:

> Thence we conclude that a balance-clock [and not a pendulum clock, according to a footnote added in 1923] at the equator must go more slowly, by a very small amount, than a precisely similar clock situated at one of the poles under otherwise identical conditions.[110]

This experiment has been done, using sufficiently accurate modern electronic clocks, and there is no detectable difference between the clocks.[111] Einstein's error was to ignore gravity. While there is more motion at the equator, there is also more gravitational potential, and the effects on time cancel. Einstein soon realized his error, and omitted the argument from a 1907 review paper that included the rest of his 1905 paper.[112] The sea-level acceleration of gravity is the same all over the Earth, even after the rotation of the Earth is taken into account, and clocks all run at the same rate at sea level.

Causality and relativity

One of Newton's great accomplishments was to show that an inverse square law of gravity leads to elliptical orbits. The German astronomer Johannes Kepler had already modeled the planetary orbits as ellipses, and others had proposed that an inverse square force law might be keeping the planets in those orbits.[113] The idea was that any two masses exert a gravitational force on each other. The amount of the force is equal to the product of the masses divided by the square of the distance. Newton's big breakthrough was to show that the force law causes elliptical orbits and the rest of Kepler's laws for planetary motion.

Newton's theory was one of the great scientific achievements of all time. But it assumed a somewhat unsettling notion of causality. It taught that gravity acts at a distance.

Our everyday notion of causality is that nothing ever really happens instantaneously. An event A can only cause an event B if A occurs before B, and there is enough time for the effects of A to get to B. For example, an ocean tsunami might be caused by an earthquake 500 miles away and hour earlier. The one exception is light, which seems instantaneous, but which travels at finite speed.

Newtonian gravity is instantaneous. According to theory, two planets can be millions of miles apart, and the force between them is immediately given by Newton's law. This was called *action-at-a-distance,* and it was hard to reconcile with intuitive notions of causality. Even Newton admitted that it was "philosophically absurd", but nobody could figure out anything better.

Consider an analogy to the flow of water. Rainwater runs off into streams, and then flows into rivers and eventually into the ocean. A naïve observer might assume that the water is attracted to the sea somehow. But that would be action-at-a-distance, and we know that the world does not work that way. Water just flows downhill, and it does not know where it is going. Relativity would eventually prove that gravity is similar. Gravity is propagated in waves, and it does not know where it is going until it gets there.

The French mathematician Pierre-Simon Laplace tried to address this problem in 1805. The Earth is 90 million miles from the Sun, and it takes sunlight about 8 minutes to get here. He figured that if the Sun's gravity also took 8 minutes to get here, then the Sun would be pulling on where the Earth was 8 minutes ago. That 8 minutes is enough to completely wreck Newton's proof of elliptical orbits. Laplace concluded that the Sun's gravity must either get here instantaneously or be transmitted a whole lot faster than light. He proved that action-at-a-distance was essential to classical mechanics.

When electromagnetism was being developed in the 1800s, there was some controversy over whether to use action-at-a-distance. Electromagnetic forces obey inverse-square laws similar to gravity, and physicists tried to develop action-at-a-distance theories for electromagnetism. Ultimately, Maxwell's theory won out over its rivals, and his equations implied that electromagnetism propagates in waves at the speed of light, and not instantaneously.

Special relativity gave causality a whole new meaning. Without absolute time, there is no way to even talk about some event instantaneously affecting distant events. Moving clocks run at different rates and cannot be synchronized. There can be no action-at-a-distance as in Newtonian gravity.

Special relativity teaches that two events are causally related if you could send a light beam, or some slower signal, from one event to the other. Physically, this means that the second event is at a later time than the first event, but not too far away for light to get there in the time difference. Mathematically, this happens when the value of the spacetime metric is nonpositive. If the metric is zero, then one event could send a light beam to the other event. If the metric is negative, or equivalently, if the proper time is positive, then one event could affect the other event with slower-than-light processes. The proper time expresses how much time is available for a causal interaction.

It takes light a little over a second to get from the Earth to the Moon, so any event on the Earth is causally related to any event more than two seconds later on the Moon. You might possibly cause something to happen on the Moon by shining a flashlight on it. Light travels very fast and you aren't likely to have much effect on the Moon, but relativistic causality is much more intuitively acceptable than the idea that you could instantaneously cause events on the Moon or farther away.

Going faster than light is like going backwards in time. There is a lot of science fiction that says that time travel might be possible as long as you do not do something stupid like killing your grandfather before you are born. These time travel paradoxes make it impossible to go backwards in time when the theory respects causality. Causality forbids travel or communication faster than light.

Poincare's popular 1902 book explained that the theory of mechanics is based on equations that relate current events to nearby events at immediately preceding moments.[114] This was a way of saying that the mathematics of causality is written with differential equations. If there were some sort of action-at-a-distance, then it would be contrary to the methods of ordinary mechanics. He said that "the aether was invented to escape this breaking down of the laws of general mechanics." He goes on to explain that he thinks that all of the attempts to detect the aether will fail. In short, the aether was a way of understanding causality, and was not some measurable substance. A distant star emits light into the aether, and the light propagates through the aether until it gets to us.

Poincare wrote a textbook in electromagnetism, and an English translation was published in 1904.[115] Instead of being a mathematical analysis of Maxwell's equations, as you might expect, it was an explanation of the physical aspects of Maxwell's theory without the equations. The translator's preface noted that the book was attempting to do what Hertz said was impossible.

Poincare explains why Maxwell's theory was preferred over previous theories. He explains how electric currents induce magnetic fields and vice versa. He said that the old theory (before Maxwell) predicts that "the propagation of inductive effects should be instantaneous". He explains how this makes understanding causality difficult, and that the induction propagates at the speed of light in Maxwell's theory. He then describes the experiments that showed that the induction really does propagate at the speed of light. He says that these were the crucial experiments that demonstrated that Maxwell's theory is true.[116]

Maxwell's theory was the first truly relativistic theory. Earlier theories of mechanics by Newton and others could use different frames of reference, but they needed action-at-a-distance for gravity, so they were not relativistic. Maxwell's theory was consistent with relativistic causality. No changes to the theory were needed when the Lorentz transformations were discovered later.

Poincare spoke in 1904 of having to valiantly defend the principle of relativity, and said:

> What would happen if we could communicate by signals other than those of light, the velocity of propagation of which differed from that of light? ... And are such signals inconceivable, if we take the view of Laplace, that universal gravitation is transmitted with a velocity a million times as great as that of light?

There are even simpler examples involving rigid bodies. If you push a stick, then you imagine that the other end moves simultaneously as you push it. A very long rigid stick would allow communication faster than light. Laue pointed out in 1911 that relativity requires that rigid bodies are not really rigid. Einstein had come to the same conclusion.

Poincare was the first to realize in 1905 that nothing was going to go faster than light, not even gravity, because that is the nature of space and time. He figured that Laplace's 1805 argument had to be wrong. Gravity could propagate at the speed of light, but no faster. In Poincare's words, the gravity propagation speed must be a function of the aether. While he had previously used the aether as a figure of speech for the kinship between all optical phenomena, his 1905 paper expanded use of the term for all physical phenomena, including gravity. This was his only mention of the aether in the paper, except to point out again that all attempts to detect the aether had failed. Under this assumption and special relativity, he discovered that Laplace's effect cancels out to a first approximation, and Earth's orbit is about the same as under relativistic gravity waves, as under Newtonian gravity. Thus, relativistic causality did not violate celestial mechanics after all.

Poincare's paper makes an analogy to Copernicus proposing that the Earth went around the Sun. The Copernican model had the advantage that it connected certain properties of the planets that otherwise seemed coincidental. Likewise, Poincare's view was that the discovery of relativity theory shows that both electromagnetism and gravity propagate at the speed of light because of an underlying property of spacetime. Poincare distinguishes his view from Lorentz's view, which was that relativity was an electromagnetic effect. According to Lorentz's theory, the transformations of space and time were accompanied by similar-looking transformations of electromagnetic variables. The similarity was coincidental until Poincare announced his spacetime theory. Einstein's approach was like Lorentz's in that the electromagnetic field transformations could not be derived from the spacetime transformations, and he did not adopt Poincare's spacetime view until years later.

The mysteries of Newtonian action-at-a-distance could be resolved if gravitation theory could be replaced by a theory with a Lorentz group symmetry, like Maxwell's equations. Poincare proposed a couple of possible such theories in 1905, and proposed that gravity is propagated by gravity waves just as electromagnetism is propagated by electromagnetic waves. His formulas for gravity were not quite right, but he was absolutely correct about causality, covariance, and gravity waves.

The planet Mercury is the planet most affected by the finite speed of gravity. It is closest to the Sun, moves the fastest, and has the most elliptical orbit. The finite speed of gravity causes Mercury's orbit to not quite close up on itself, and the point where it is closest to the Sun, called the *perihelion*, changes a tiny bit with each orbit. The change is so small that it is only noticeable after a few centuries. The advance of Mercury's perihelion was noticed in the 1800s, using data going back to Tycho, and in 1859 it was shown that most, but not all, of the effect can be attributed to the planet Jupiter dragging Mercury with its gravitational pull. It turned out that there was an unexplained residual effect. It was so tiny that it would take three million years for Mercury's perihelion to make a complete revolution around the Sun. It could be partially explained by the finite speed of gravity, as Poincare announced in his 1908 book. The Dutch cosmologist Willem de Sitter published an analysis of how different relativistic gravity theories might explain Mercury's orbit in 1911.[117] Einstein explained it more completely in 1915, using Grossmann's relativistic theory of gravity.[118]

Understanding causality was a major motivation for Poincare to create relativity theory. Einstein's 1905 paper said nothing about causality. He began to consider causal issues in 1907. Relativistic causality comes from Poincare and others, not Einstein.

Explaining the physics

Poincare was a mathematician, and it is sometimes presumed that he understood the mathematics of relativity and not the physics. Lorentz was a physicist, but he is also accused of treating his transformation as a mathematical construct that is not necessarily observable. Others say he gave an electromagnetic theory of relativity when it should really be just measurement theory. Einstein is popularly credited with understanding the physics properly.

In fact, Einstein's famous 1905 paper does not really explain the physics of relativity very well at all. There is barely any mention of any experiments to support the theory. The paper is based on postulates, not experiments. For his first postulate, the relativity principle, he does mention "unsuccessful attempts to discover any motion of the earth", but he does not mention the most prominent and important such attempt, the Michelson-Morley experiment. In his later life, he denied that the experiment was even a consideration for him. His second postulate is about the speed of light, and he does not give any physical reason or evidence for believing that the speed is constant.

To Lorentz and Poincare, the experimental foundations for relativity were crucial. The whole point of the theory was to reconcile electromagnetism with the Michelson-Morley experiment.

Einstein gives the formulas for how motion makes a rigid measuring rod contract and for how time slows down, but says nothing about what is really going on physically. Lorentz attempted to explain these effects in terms of electromagnetism. It was known from Maxwell's equations that motion has a contracting effect on an electric field. If that measuring rod is held rigid by electromagnetic forces, then perhaps the rod shortens because of changes in those forces. Likewise, electromagnetic effects might also explain the slowdown of clocks. Poincare explained relativity as being fundamental to the nature of measuring space and time, although he conceded the possibility that "everything in the universe would be of electromagnetic origin", as Lorentz apparently believed.

Einstein's approach is sometimes called *operationalism*. He talks about how an observer makes measurements with meter sticks and clocks, without attempting to describe the physics of what is really going on. He was later to repudiate this view when Bohr interpreted quantum mechanics in terms of observations, and when Einstein himself developed unified field theories with no observable properties.

While Einstein did not give a physical explanation of the rods contracting, his belief was that such an explanation would have been desirable. In his 1916 relativity book he complained that the contraction was "not justifiable by any electrodynamical facts".[119] It was an unexplained consequence of his postulates. He sometimes talked about the lengths of measuring rods being determined by electrodynamic effects. He later wrote that he searched for such a constructive presentation of relativity and failed.[120] Pauli later agreed on the need for such an explanation:

> Should one, then, completely abandon any attempt to explain the Lorentz contraction atomistically? We think that the answer to this question should be No. The contraction of a measuring rod is not an elementary but a very complicated process. It would not take place except for the covariance with respect to the Lorentz group of the basic equations of electron theory, as well as of those laws, as yet unknown to us, which determine the cohesion of the electron itself.[121]

Today the preferred view is that relativity is a property of spacetime itself, and Einstein adopted that view after Minkowski popularized it in 1908. The electromagnetic view is tenable also, as it was later shown that quantum field theory could give what FitzGerald and Lorentz hypothesized. That is, it is possible to understand moving yardsticks as getting shorter because the Lorentz-transformed electromagnetic fields are moving their atoms closer together.[122]

By explaining the physics of length contractions in terms of the electrodynamics of rigid bodies, it is sometimes said that Lorentz's relativity was a conspiracy of dynamical effects. That is, the motion causes a deformation of the electromagnetic field, which in turn causes a contraction of rigid bodies. The contraction miraculously turns out to be exactly what is needed to explain Michelson-Morley, and make it appear that the speed of light is constant. Thus the constancy of light speed appears to be a lucky coincidence in Lorentz's theory, whereas it is a postulate in Einstein's. Some say that it was an advantage of Einstein's approach that he did not try to explain the dynamical effects, and did not rely on coincidences.

But the coincidence is not really a coincidence. Explaining Michelson-Morley was Lorentz's motivation for developing his transformation, and likewise for FitzGerald and Larmor. Voigt had other motives for finding similar transformations, but he was also looking at symmetries that would preserve the speed of light. Lorentz's theory was no more coincidental than Einstein's. Lorentz's theory is richer because he gives a valid physical explanation for the length contraction.

Minkowski regarded Lorentz's theory as being the same as Einstein's. Here is how Minkowski compared the theory to his own spacetime geometry view:

> But all efforts directed towards this object [to detect the motion of the Earth], and even the celebrated interference-experiment of Michelson have given negative results. In order to supply an explanation for this result, H. A. Lorentz formed a hypothesis ... According to Lorentz every substance shall suffer a contraction [formula omitted] in length, in the direction of its motion.
>
> This hypothesis sounds rather fantastical. For the contraction is not to be thought of as a consequence of the resistance of aether, but purely as a gift from the skies, as a sort of condition always accompanying a state of motion.
>
> I shall show in our figure that Lorentz's hypothesis is fully equivalent to the new conceptions about time and space. Thereby it may appear more intelligible.[123]

In other words, the Lorentz transformation was some sort of "gift from the skies" to keep the speed of light constant for all observers. Lorentz's hypothesis was a property of motion, not the aether.

Poincare's 1905 view was also that the symmetries of spacetime give an alternative explanation equivalent to Lorentz's. He also theorized about the shape of an electron, and how it might be deformed when it moves. Negative charges repel other negative charges, so it seemed that there must be some forces holding the electron charge together. These forces are sometimes called the *Poincare stresses*, or *pressure*. He was concerned about the self-energy of the electron if its volume changes.

The analysis of shape gives a way of understanding a relativistic electron. The pressure keeps the charge together, in a small sphere, and the self-energy of the charge gives the electron mass. The motion causes an additional pressure in the direction of the motion, and contracts the electron's shape to an ellipsoid. You could imagine the contraction being caused by resistance from the aether, although Poincare did not actually say that. The smaller volume makes the electron charge more concentrated, and hence having more self-energy and more mass. Thus Poincare was able to relate the electron shape to the length contraction and mass increase and thereby give a physical interpretation of the Lorentz transformation.

Einstein's 1916 relativity book said, "The general theory of relativity renders it likely that the electrical masses of an electron are held together by gravitational forces." No one believes that today.

Nowadays, the electron is usually considered a zero-volume point particle with wave properties, and the self-energy is calculated by quantum mechanics in other ways. Many physicists today believe that electrons have some other shape, in part because of its paradoxical self-energy. The Poincare pressure did not turn out to be as important as the rest of his relativity theory, but it is still sometimes mentioned in textbooks. The main use of this analysis was that Poincare was able to use it to distinguish his theory from alternatives. By that time there were three competing theories for how mass increases with velocities, with numerically different predictions. Poincare argued that the other two were wrong.

Einstein's defenders sometimes also argue that Poincare was just a mathematician who corrected Lorentz's mathematical errors without understanding the physics, while Einstein understood the physics. But as you can see above, Poincare addresses the physics more directly than Einstein, and has the courage to say that the other physicists' theories were wrong.

Historian Arthur I. Miller studied the writings of Poincare and Einstein, and argued that the chief difference was that Poincare was willing to explicitly admit that experiments could prove him wrong, and Einstein was not. He points out that Poincare mentioned Kaufmann's experiments on relativistic mass that seemed to contradict his theory.[124] Kaufmann published an experiment in Nov. 1905, saying, "the results are not compatible with the Lorentz-Einstein fundamental assumptions." Lorentz and Poincare wrote letters of concern to each other. (More careful experiments later showed that electron inertia does increase with velocity as Lorentz and Poincare predicted.) Miller said that Einstein should get the credit for special relativity because he was willing to believe it regardless of the evidence. Einstein biographers Pais and Walter Isaacson go even farther, and argue bizarrely that Poincare never understood the basis of special relativity.[125] But in fact Poincare's understanding was superior to Einstein's on every single facet. Pais was an accomplished physicist, but he showed his own lack of understanding when he said, "Einstein proved the Lorentz covariance of the Maxwell-Lorentz equations".[126] Einstein postulated something similar to covariance, namely Lorentz's theorem, and did not prove it. It is to Poincare's credit that he was willing to consider contrary evidence.

Reality of the transformations

When Einstein starting trying to distinguish his theory from Lorentz's in 1907, he said that a key insight was to realize that Lorentz's local time was really time. It was as if Einstein answered the question in the popular 1969 Chicago song, *Does Anybody Really Know What Time It Is?* Minkowski echoed this saying that Einstein was able "to perceive clearly that the time of an electron is as good as the time of any other electron". It is often said that Einstein's greatest insight was to realize that the Lorentz transformations were real.

The transformations are certainly real, but it is not so easy to say what that means. Physics is filled with sophisticated mathematical theories, and there is no rule for saying when a variable is a real physical entity, or when it is just a clever mathematical shortcut. For example, Maxwell's electromagnetic fields are usually regarded as real, but they can also be mathematical constructs for calculating the effect of electric charges on other charges. Debates about the reality of astronomical models go back millennia.

Consider temperature, as an everyday example of a physical variable. It is easily measured by thermometers, and is obviously useful. But it is not so easy to say that it is real. For a long time, people knew that heat and cold were opposites, but it was not clear whether they were some sort of substances or what. It was not until the 1800s that physicists discovered a theory of thermodynamics, in which heat was a form of energy and cold was the absence of heat.

Thermodynamics teaches that heat is the energy of molecular motion. A moving object has energy called kinetic (or mechanical) energy. Even when an object is at rest, its atoms and molecules are always vibrating, and that vibration is the heat energy. Only at a temperature of absolute zero does the vibration stop and there is no heat energy. When you apply the disc brakes on your car, you convert kinetic energy into heat energy. A basic fact of thermodynamics is that the conversion is irreversible.

All that does not tell us whether temperature is real. It feels real, but it can also be seen as just a mathematical shortcut for describing molecular motion. A realist would argue that temperature has some objective existence independent of our observations.

Relativity teaches that a moving object has a length contraction, time dilation, and mass increase. Lorentz certainly believed that these transformations were real in the sense that the Michelson-Morley apparatus was affected, as that was how he explained the experiment. Einstein believed that it was real in the sense that measuring rods and clocks would be affected. Lorentz's and Einstein's views were so nearly identical that there is no agreement about which was more realist. Some argue that Einstein, and not Lorentz, said that moving clocks really slow down. For example, Galison wrote a whole book on Einstein's clocks, and argued that Lorentz had only a "purely mathematical idea of time".[127] English Astrophysicist Stephen Hawking blames Poincare for having "regarded this problem as mathematical", and Lorentz for believing that the clocks really slowed down. He praises Einstein for avoiding such an explanations and having the view that the measurement of time depended on the observer.[128] Either way, these distinctions were developed decades later. No one saw any such difference between Lorentz's and Einstein's views in 1905, and subsequent papers referred to the "Lorentz-Einstein" theory.[129]

New York Times science editor Dennis Overbye wrote an Einstein biography that blames Lorentz for saying that the transformations were real, and adds:

> In a way, the message of relativity theory was that physics was not about real objects, rather, it concerned the *measurement* of real objects.[130]

The book suggests that Einstein had a different meaning from Lorentz, but it admits that Einstein made "no such declaration of grandeur, of course". Only Poincare did that.

It is possible to take a view of relativity that nothing physical is really going on. One could argue that objects do not really contract, but just appear to contract because of an illusion stemming from the choice of a reference frame. The contraction could be considered just an artifact of how we choose to define measurements. The apparent changes in time and mass could also be considered just changes in how we do measurements. This is the 4-dimensional geometric view.

If it were possible to somehow attach a clock to an electron, then perhaps that could be used to define the time of an electron. But that is impossible. All we can do is to infer a notion of time based on electromagnetic theory and decide how to interpret measurements. There are other strange theories of electrons, besides that they have their own internal time. Dirac proposed that we live in an invisible sea of electrons, and that positrons are just bubbles in that sea. Feynman proposed that positrons are electrons going backwards in time. His quantum field theory is based on electrons having infinite mass and charge, but measurements are finite because of virtual particles in the aether. And of course there is also the view that there are no particles, just waves.

Mathematicians prefer the geometric view. Without some preferred frame of reference, it is impossible to say that anything is moving, and hence impossible to say that any length is contracted or any clock is slowed. You can only say that objects look that way when viewed from some other frame with some relative motion. The contraction is thus an illusion based on how the object is viewed.

Two meter sticks are not so easily compared when there is relative motion, and two clocks are not either. We have adopted conventions for making those comparisons, and there are some funny outcomes in what seems like our 3-dimensional world. If we could see all four dimensions with its non-Euclidean geometry, the outcomes would not seem so funny.

Poincare announced this geometric view in 1905, and proposed that relativity was "something which would be due to our methods of measurement." Relativity is often described today as a theory of measurement. Sometimes the effects on rods and clocks are described as "apparent" to emphasize that the theory is only saying how they appear to moving observers.

Poincare wrote in 1907, "And yet we have no means of knowing whether this deformation is real."[131] He was not doubting the Lorentz transformation. He was pointing out the difficulties in testing our assumptions about space and time. He also said, "Time itself must be profoundly modified."[132] He did not just say that measurements appear to be modified. Time is modified.

The trouble with the geometric view is that it suggests that the relativistic effects are not real, even though they seem as physical as anything else. It is tough to argue that magnetism is just an illusion.

The Yugoslav mathematician Vladimir Varicak wrote a couple of papers in 1910 on non-Euclidean geometry interpretations of relativity. He even went so far as to argue that the length contraction was a psychological effect. That was too much for Einstein, and he wrote a rebuttal.[133]

The twin paradox shows that the Lorentz transformation is not just a psychological effect. If two identical twin boys are born on Earth, and one goes on a futuristic rocket ship trip going close to the speed of light, then he can return to Earth much younger than his Earth-bound brother. Relativity teaches that one twin can be 50 years older than the other when they are reunited, just as if one had been frozen for 50 years.

The easiest way to think about the twin paradox is to say that time slows down on the rocket ship, but it is not so clear that view is correct. The brother on the rocket ship will deny that time slowed down for him, as he will not perceive any slowness. One interpretation is to say that they aged at the same rate but took different paths through spacetime. By that interpretation, no clocks really slow down, and it only seems like a paradox because we are not used to the 4-dimensional non-Euclidean geometry.

There is no simple answer to the question of whether the Lorentz transformations are real. The effects on measurements are real, and that is what physics can be sure about.

Relativistic action

The most widely known theory of physics is Newtonian physics, developed by Isaac Newton in the late 1600s. It is called *classical mechanics*, it is is usually taught in terms of forces on particles. Newton's law says that force equals mass times acceleration. Different formulations of mechanics were discovered by the Italian-born mathematician Joseph-Louis Lagrange in 1788 and the Irish mathematician William Rowan Hamilton in 1833. They are called *Lagrangian* and *Hamiltonian* mechanics. These three systems are conceptually very different, but they can be used to solve the same problems of classical mechanics. As an example of the conceptual differences, Newtonian and Hamiltonian mechanics define the total energy as the kinetic energy plus the potential energy. Lagrangian mechanics subtracts them instead of adding them. When that difference is multiplied by time, it is called the *action*.

It makes sense to add the different forms of energy, because the total energy is conserved. Potential energy can be converted to kinetic energy, and back to potential energy (plus heat energy from friction, possibly), and the total energy remains the same throughout. It seems to be conceptually erroneous to subtract energies. Energies are meant to be added, not subtracted. But Lagrangian mechanics is based on subtracting the energies.

The action is mysterious, and hard to explain physically. But somehow minimizing the action has all the same observational consequences as Newtonian and Hamiltonian mechanics so it is certainly not wrong. Even today, there is no general agreement about which of these three systems is best, and the action is often used in modern quantum field theories.

The units of action are the same as for Planck's constant, the fundamental constant that Planck discovered in 1900 that became so essential to quantum mechanics. That constant defines how light is quantized into photons. It relates position and momentum in much the same way as the constant speed of light relates space and time in relativity. This quantum of action is one of the three or so most important constants in all of physics.

Lagrange discovered the *principle of least action*, whereby he could solve mechanics problems by minimizing the action. Lagrange was not the first to discover the principle, as there was once a hot dispute over who deserved credit for it. The Berlin Academy of Science once held a trial of sorts over claims that it was first used in private letters from the German mathematician Gottfried Wilhelm von Leibniz. Leibniz was a rival of Newton's, and had discovered infinitesimal calculus before Newton published his ideas, and there was a heated debate over who deserves the credit for that. In 1752 the Academy decided, unfairly it seems, that the Leibniz letter was a forgery. Apparently priority disputes were taken quite seriously back then.

Lagrange and Hamilton developed the principle into full-fledged alternatives to Newtonian mechanics. These alternative theories were not necessarily any better, but they led to a much improved conceptual understanding of mechanics, and they were simpler for certain types of problems. The real payoff to these alternative systems of mechanics came in the 20th century, when Newtonian mechanics got replaced by quantum mechanics. It was often easier to define a Lagrangian or Hamiltonian function than to make sense out of Newtonian forces on particles.

When Poincare gave his 1904 St. Louis lecture on the principles of mathematical physics, he listed the principle of least action as one of the five or six most important general principles. Others included conservation of energy and the relativity principle. When he gave a derivation of Lorentz transformations in 1905, he used the principle of least action.

More importantly, Poincare formulated a *relativistic action* in terms of an electromagnetic field. He showed that certain combinations of the electric and magnetic fields were unchanged by Lorentz transformations, and he defined the action to be a certain function of these. Electromagnetism could then be understood as what happens when you minimize the action, just as in Lagrangian mechanics. Applying the principle of least action gave a new derivation of Maxwell's equations. The German physicist Karl Schwarzschild found a similar action in 1903. Poincare proved that the Lorentz transformations were symmetries of the action, and hence he had a second proof that they were symmetries of Maxwell's equations. He thus gave a new formulation of electromagnetism that more obviously satisfied the relativity principle, and that made the Lorentz group invariance more transparent.

While physical theories are usually given in terms of equations of motion, they can also be given by specifying an action function, and allowing the consequences to be deduced by applying Lagrangian mechanics. Poincare thus gave a clever and concise way of formulating a relativistic theory. Just define a Lorentz-invariant action and all else will follow.

Pauli later wrote, in his 1921 book on relativity, "Mathematically speaking, therefore, the special theory of relativity is the theory of invariants of the Lorentz group." It was only Poincare who explicitly sought and found the invariants of the Lorentz group in 1905, as a way of formulating physical theories.

Decades later, Poincare's approach became crucial for quantum field theory. The action was used for quantum electrodynamics, and then later for other quantum field theories. A relativistic action was also used to derive a relativistic theory of gravity. A lot of 20th century physics was inspired by formulating an action that obeys the appropriate symmetry principles.

3. The great Einstein myth

The history of relativity does not explain why Einstein came to be regarded as such a great genius. The facts in this book have been known for a century. His fame was based on factors other than the originality of his papers.

Why Einstein is idolized

The New York Times newspaper has been idolizing Einstein since 1919, when its front page headlines read:

> LIGHTS ALL ASKEW IN THE HEAVENS; Men of Science More or Less Agog Over Results of Eclipse Observations. EINSTEIN THEORY TRIUMPHS Stars Not Where They Seemed or Were Calculated to be, but Nobody Need Worry. A BOOK FOR 12 WISE MEN No More in All the World Could Comprehend It, Said Einstein When His Daring Publishers Accepted It.

Other newspapers were even more gushing. The London Times headline said:

> REVOLUTION IN SCIENCE – New Theory of the Universe – Newtonian Ideas Overthrown – Momentous Pronouncement – Space 'Warped'[134]

The British Royal Society had just discussed the eclipse observations, and announced that "Einstein's reasoning … is the result of one of the highest achievements of human thought."

A physicist complained that "The skeptics were led by the New York Times."[135] The newspaper was more gullible than skeptical. A follow-up article said:

> … Einstein is a great Swiss mathematician and physicist, who holds a professorship in Germany at present. His trump card

> was the "principle of relativity," which is the new theory that has set the world agog. "Lorentz's theory is a side issue," claimed Einstein. The fundamental thing is the principle of relativity which expressed simply is this: The universe is so constituted that it is impossible to detect the absolute velocity of the motion of any body through space.

Several times later, the newspaper had breathless stories about new theories that Einstein supposedly discovered. Even when he had nothing, it ran the headline, "Einstein baffled by cosmos riddle".[136] It never explained where he got that "trump card".

Einstein told the Times in 1921 that opposition to his relativity theory was entirely anti Semitic.[137] Some of the anti-relativists did complain about Jewish conspiracies, but the German physicist Ernst Gehrcke was more typical, and he was concerned about the future of science. New Scientist magazine reports:

> The increasing role played by advanced mathematics seemed to disconnect physics from reality. "Mathematics is the science of the imaginable, but natural science is the science of the real," Gehrcke stated in 1921. Engineer Eyvind Heidenreich, who found relativity incomprehensible, went further: "This is not science. On the contrary, it is a new brand of metaphysics."
>
> The Academy of Nations therefore saw itself as directed not only against the theory of relativity, but also towards the salvation of what it considered to be real science. Gehrcke insisted that the Academy "must become an alliance of truth".[138]

By 1922, the newspaper was overwhelmed with relativity books to review, and it complained that relativity had already become a "cult" that "mathematicians and metaphysicians" had nearly all joined.[139] Time magazine later said that "Einstein is the only scientist who has become a cult figure, even among scientists."[140]

The anti-relativity movement reached its peak in 1931 with publication of a German book, *Hundred Authors Against Einstein*. It had a collection of essays by relativity skeptics. The arguments were not very scientific, but were not anti-Semitic either.

The biggest newspaper headlines were for Einstein's work on gravity, but he is mainly idolized for his paper on special relativity. Physicist Brian Cox recently said this, while promoting his latest book on Einstein:

> Relativity is the basis on which all of our understanding of modern physics rests. So without relativity, we would not understand how transistors work, how cell phones work, [and] we wouldn't understand the universe at all without relativity. It is the foundation on which [all modern science] rests.

What he means by this is that much of 20th century physics was developed by finding theories that obeyed the symmetries of spacetime, such as Lorentz transformations. He is talking about special relativity, not gravity. And Einstein is often given the entire credit for special relativity, as if no one else contributed anything, as exemplified by Princeton physicist Valentine Bargmann, who wrote:

> The first paper on the special theory of relativity (written in 1905, when Einstein was an employee of the Swiss Patent Office at Berne) presents the theory already in final form.[141]

Even Einstein's mistakes do not detract from his reputation. A recent book catalogued dozens of them, both large and small. The book has a postmortem saying:

> Einstein made so many mistakes in his scientific work that it is hard to keep track of them. ... It's a bad record. Despite all of these mistakes, Einstein was unquestionably the greatest physicist of the twentieth century, and he was the second-greatest physicist ever, outranked only by Newton.[142]

Much of Einstein's 1919 fame came from the British astrophysicist Arthur Eddington. He was a Quaker and a World War I conscientious objector, and he managed to get a project to take solar eclipse photographs instead of having to fulfill his military obligations. He was the one who announced that Einstein had been proven correct, even though his data was not so decisive. He thought that a British project to verify a German theory would be good for European peace, as the British had just defeated Germany in the war.[143] He wrote a 1923 book on relativity and became a big Einstein promoter. After describing Lorentz transformations using Poincare's imaginary time trick (without mentioning Poincare), he wrote:

> Historically this transformation was first obtained for the particular case of electromagnetic equations. Its more general character was pointed out by Einstein in 1905.[144]

But this is completely false. The FitzGerald contraction was first obtained as a consequence of the constancy of the speed of light, just as Einstein did 16 years later. Poincare pointed out its more general character in 1905, but Einstein did not.

The English mathematician Edmund T. Whittaker wrote a 1953 treatise on electromagnetism that shocked people by how little credit it gave to Einstein for special relativity:

> Einstein published a paper which set forth the relativity theory of Poincaré and Lorentz with some amplifications, and which attracted much attention.[145]

Many physicists overreacted to this, and rushed to Einstein's defense. Einstein's friend and colleague Max Born had even tried to persuade Whittaker not to publish this opinion. Born learned relativity as a student from Hilbert and Minkowski, and later wrote, "Einstein's paper was a revelation to me which had a stronger influence on my thinking than any other scientific experience". He also wrote a popular book on special relativity titled, "Einstein's Theory of Relativity". But he later admitted that Einstein's 1905 reasoning was the same as Poincare's, and that it was possible that Poincare had all of special relativity before Einstein.[146] Actually Einstein's 1905 paper really just sets forth the Lorentz-Poincare relativity theory of 1900, and does not have any of Poincare's 1905 theory, such as 4-dimensional spacetime, symmetry group, metric tensor, electromagnetic covariance, relativistic action, and gravity.

The New York Times came to Einstein's defense in a 2001 book review:

> This Arthur I. Miller is the one who finally disposed of Edmund Whittaker's claim that Poincaré was the true discoverer of special relativity.[147]

Miller's analysis was that Poincare and Einstein invented versions of special relativity that were mathematically and observationally identical. (He ignored Einstein's low-velocity approximations in the forces on electrons.) He also says that Einstein could have been substantially influenced by Poincare, but he stops short of accusing Einstein of plagiarism. Miller argues that Poincare considered his theory as an improved version of Lorentz's electron theory, and Einstein did not cite any of his sources, so "the honors for special relativity go to Einstein, alone".[148]

If that sounds backwards, it is. The reasons for idolizing Einstein *are* backwards. Nobody seriously favors Einstein because of the mathematics or the physics of special relativity. They like Einstein because of the boldness of the position that he symbolized. Lorentz, Poincare, and the other relativity pioneers seem timid by comparison, as they generously credited others and failed to demand personal recognition. The title to Lorentz's big 1895 book started with the German word *Vorsuch* for "attempt", and Poincare's big 1905 paper ended the preface by modestly saying, "I, too, have not hesitated to publish these few partial results, even if at this very moment the discovery of magneto-cathode rays seems to threaten the entire theory."

The San Francisco 49ers football team used to have two excellent quarterbacks. The hero was Joe Montana, and he led the team to some championships. The backup quarterback was Steve Young. Young was a team player, and he seemed content to sit on the bench while Montana played. For that, he lost the respect of the local fans. Even when Young eventually replaced Montana and led the team to a championship, he still did not gain the fans' respect. A true winner like Montana would have refused to sit on the bench, they said. But Young was an outstanding quarterback by any objective measure. He currently holds the highest career NFL passing rating and several other records.

Likewise, it appears that Poincare is not respected by some people because he was willing to share credit for his theory. The glory in physics now goes to the wild-eyed egomaniac who claims to lead a revolution. For example, Holton accuses Poincare of gradualism, and trashes him for relating his work to older theories and for not recognizing Einstein as a revolutionary.[149]

Once convinced that special relativity was revolutionary, many historians refused to believe that it could have come from someone like Poincare who was already an established intellectual. No well-respected scientist would ever be so bold as to risk his reputation on such an outlandish theory. But they are wrong.

Some historians give other reasons for preferring to favor Einstein for the invention of special relativity. The most common reason given is that they prefer Einstein's terminology. They prefer Einstein's use of the term "stationary system" over Lorentz's use of the term "aether". They acknowledge that the difference had no mathematical or observational significance, but that only adds to Einstein's genius, they say. They argue that only Einstein staked out a conceptual position that was impervious to any scientific test, and for that he was truly great.

Sometimes a different formulation of an equivalent theory can lead to useful results. But Einstein's formulation of special relativity had only a pedagogic value. It was Poincare's formulation that was used in subsequent physics. Poincare described spacetime with a metric tensor, and as a homogeneous space. Poincare explained that the laws of physics must therefore be invariant under the Lorentz group, and the larger group that is now known as the *Poincare group*.[150]

Popular accounts of Einstein's relativity today are often dumbed-down with counterintuitive stories about trains, clocks, and lightning bolts. Not everyone agrees with the pedagogic value of this approach. One physicist recently wrote that relativity is only easy once you learn about the 4-dimensional geometry:

> Relativity is "hard" because there's like one million books full of confusing stories about spaceships and lasers and somebody observing somebody's something, which is all completely irrelevant decoration. As a teenager I read a whole stack of these books and failed to make much sense out of them because one starts asking all sorts of questions about the construction of clocks and what it means to actually "see" something etc. Then, hallelujah, somebody handed me a book in which it said the Poincaré-group is the symmetry group of Minkowski-space.[151]

The metric turned out to be the fundamental object of study to create a relativistic theory of gravity. Modern theories of the electron, photon, and other elementary particles were all developed with Poincare group invariance being a guiding principle. Poincare was the first to search for such theories.

Many physicists preferred Einstein's presentation because it was more elementary, and much of it could be read without understanding either symmetry groups or Maxwell's equations. Einstein gives an illusion of having proved something rigorously, and doing it with physically plausible arguments. Even today, many physicists mistakenly say that Einstein proved the Lorentz invariance of Maxwell's equations.

Lorentz later wrote that, "Einstein simply postulates what we have deduced".[152] He meant that he (and Poincare) had deduced relativity from electromagnetic theory and experiments like Michelson-Morley. Einstein just postulated that Maxwell's equations were relativistic, and discussed the consequences. Einstein's first relativity postulate was essentially equivalent to what was previously known as Lorentz's theorem of the corresponding states. It must have seemed like a cheat to Lorentz. It is nice to have an explanation of a theory based on simple assumptions, but it would have been simpler still to assume Poincare group invariance, or to assume that the only measure of distance and time is the metric tensor.

While these historians prefer to credit Einstein, no one can dispute the fact that Poincare discovered all of the elements of special relativity with help from Lorentz and others, published them before Einstein, and developed a theory that was either identical or observationally equivalent to Einstein's.

Poincare was a real genius

Henri Poincare was a famous French intellectual. He did pioneering mathematical work in differential equations, topology, and mathematical physics. He published about 500 papers. The Hungarian Academy of Sciences called him "incontestably the foremost and most powerful researcher of the present time in the domain of mathematics and mathematical physics" when it gave him a prize in 1905.[153] He wrote well-respected essays in philosophy and other fields. He wrote several popular science books that became bestsellers. He even worked as a mining engineer. Today he is largely remembered for founding chaos theory, which explains the *butterfly effect*, and for the *Poincare Conjecture*.

The butterfly effect is the idea that some systems, like the weather on Earth, are so chaotic that very tiny changes can cause a cascade that ultimately causes a huge effect. In theory, a butterfly flapping its wings could cause a storm on the other side of the world a year later. The storm would have many other more immediate causes, of course, but without the butterfly, the storm might not have happened. In math jargon, weather is sensitive to initial conditions. Even if you could determine all the world's weather variables to ten decimal places, there would still be limits on your ability to predict future weather. Earth's weather is a chaotic system.

Poincare discovered chaos theory in celestial mechanics. He showed that if planets are orbiting each other in a plane under Newtonian gravity, then they settle down into stable long-term behavior. But in three dimensions, they can become chaotic, and the long-term behavior is nonperiodic. Here is how he described chaos in a 1908 essay:[154]

> It may happen that small differences in the initial conditions produce very great ones in the final phenomena. A small error in the former will produce an enormous error in the latter. Prediction becomes impossible.

Poincare was way ahead of his time. A 1988 best-selling book on the subject of chaos theory called it a new science.[155] His ideas also play a major role in the recent best-seller, *The Black Swan*.[156]

The Poincare Conjecture stumped mathematicians for a century. It was a geometrical problem about what kinds of 3-dimensional manifolds are possible. It said that the only way for a 3-dimensional space to close up on itself is like a sphere, or else it must have holes that are detected by the groups that he invented. The final solution several years ago is the most important mathematical work of the 21st century so far. The American journal AAAS *Science* declared the proof to be the scientific breakthrough of the year for 2006,[157] and that is the only time a mathematical work has been so honored. The solution led to a geometric classification of 3-dimensional manifolds.

Poincare invented the subject of algebraic topology with a long 1895 paper on what he called *analysis situs*. He figured out how to use algebra and group theory to characterize the topology of spaces, such as the number of holes in a surface. It was brilliant and original, but not completely rigorous and it left many loose ends. He wrote several follow-up papers and corrected some of his own errors over the next few years. He discussed 2-dimensional surfaces, as well as higher dimensional manifolds that are not so easily visualized. He had a topological condition for a manifold to be a 3-dimensional sphere. The final paper in the series was published in 1904, and it gave a counterexample – a 3-dimensional manifold that looked like a sphere but was actually tied in a knot. The paper ended with a proposal for a stronger topological condition for being a sphere, and that condition became known as the Poincare Conjecture. More modern mathematics was inspired by this problem than any other problem of the century.

Poincare was a pioneer in non-Euclidean geometry, where a space has a metric and curvature, and topology, where spaces have properties that do not depend on a metric or curvature. In the latter part of the 20th century, many relationships between curvature and topology were discovered by the Chinese-American mathematician Shing-Tung Yau and many others. He explains some of this work in a recent book, and says:

> In mathematics, no one can claim to have started anything from scratch. The idea of geometric analysis, in a sense, dates back to the nineteenth century — to the work of the French mathematician Henri Poincaré, who had in turn built on the work of Riemann and others before him.[158]

It was shown that if a certain curvature tensor were allowed to dissipate like heat, then a space could be deformed into one with more uniformly distributed curvature and possibly simpler topology. The curvature tensor was the Ricci curvature tensor, named after the Italian mathematician Gregorio Ricci-Curbastro. The Russian mathematician Grigori Perelman made a detailed analysis of how curvature flows can cause singularities, and how removing those singularities affects the topology. He was able to prove that any closed 3-dimensional manifold can be topologically decomposed into simpler manifolds with well-understood geometries. This solved the Poincare Conjecture in 2003.

Even if it turned out that Poincare's treatment of relativity were deficient in some way, he still should be credited with what he did correctly. The German physicists Heisenberg and Erwin Schroedinger are credited with inventing quantum mechanics, even though they missed the crucial idea that the theory predicts probabilities.

I am not trying to convince you that Poincare should be idolized instead of Einstein. Crediting Poincare with relativity will not change his reputation much. Most of his obituaries did not even mention relativity. A lengthy tribute by the French Academy of Sciences[159] a year after his death had just one sentence on his 1905 relativity papers. It only said that he perfected Lorentz's theory. The Astronomical Society of the Pacific gave him a medal in 1911, and suggested that he was so much smarter than everyone else that it would takes decades to understand what he had done. The presentation said:

> Poincaré's investigations have dealt chiefly with pure mathematics, with mathematical physics, and with mathematical astronomy. Very few living men of science are competent to speak

> in detail concerning the value of his work, but there can be no doubt that it is of very high rank. His comprehension of the general principles underlying these subjects is remarkable. In giving mathematical expression to these principles Poincaré possesses real genius, probably unequalled by that of any other living man.[160]

A recent book on the Poincare Conjecture only had one paragraph on his relativity work.[161] The same year he announced his new mechanics of relativity, he also published a topology paper with his famous conjecture. If he is credited for inventing special relativity, it will be considered one of his minor accomplishments. The famous British philosopher Bertrand Russell once said that Poincare was the greatest man France had ever produced.

Poincare made his share of mistakes. In 1884, the king of Sweden and Norway offered a public prize for the best solution to a problem in celestial mechanics.[162] The manuscripts were to be submitted anonymously in 1888, and judged by leading mathematicians. Poincare won the contest with the pseudonym "nothing exceeds the limits of the stars" in Latin.[163] The award generated several controversies, and no such prize was offered again. The biggest problem was that Poincare discovered an error just as the 150-page paper was being published, and it was going to be an embarrassment to everyone involved. He agreed to pay to destroy and redo the print run of the journal, and the cost exceeded the amount of the prize money that he received. The problem was that he had underestimated the possibility of chaotic orbits and he had not invented chaos theory yet. He revised his paper for publication and fixed the error.

There are different theories for why Poincare has not received more credit for relativity. One possibility is that his cousin, Raymond Poincare, was President of France from 1913 to 1920. Physics was dominated by Germans in the early 20th century, and most Germans hated the French President for political reasons related to World War I.

Einstein wrote his papers in a way that was appealing to physicists. His mathematics was elementary and his explanations were more self-contained. His reasoning was physically plausible, if not mathematically rigorous. Poincare wrote scientific papers at a higher level, and lower-level essays for the general public. If you read only his popular essays, it might be possible for you to believe that he stumbled upon some brilliant ideas without realizing the ramifications of what he was saying. But his big 1905 relativity paper has 50 pages of detailed analysis that show his understanding to be years ahead of everyone else. It was not easy to read. Advanced papers by mathematicians frequently are not easy to read. Perelman announced the proof of the Poincare Conjecture in 2003, but it was not accepted as correct until three years later. Most physicists learned Poincare's ideas from textbooks, and do not learn the source of those ideas.

I am also not arguing for a Poincare revolution instead of an Einstein revolution. My argument is that the development of science has been gradual and continuous over thousands of years, and the whole idea of scientific revolutions is mistaken. Poincare built on the theoretical and experimental work of others.

Top-down theories

Computer scientists draw a subtle distinction between *top-down* and *bottom-up* designs. A top-down design proceeds from high-level components to low-level components, and a bottom-up design does the reverse.

If you read a book top-down, you would read the title, then the table of contents, and then the text of the book. Reading it bottom-up would start with the text. A better example is writing an article. The top-down method is to write the title, then a short outline, then progressively more detailed outlines until you have a complete article. The bottom-up method would be to immediately write paragraphs of text, and then piece them together into an article. Both approaches have merit, of course.

In a popular explanation of relativity, Einstein wrote in 1919:

> We can distinguish various kinds of theories in physics. Most of them are constructive. They attempt to build up a picture of the more complex phenomena out of the materials of a relatively simple formal scheme from which they start out. …
>
> Along with this most important class of theories there exists a second, which I will call "principle-theories." These employ the analytic, not the synthetic, method. …
>
> The advantages of the constructive theory are completeness, adaptability, and clearness, those of the principle theory are

> logical perfection and security of the foundations. The theory of relativity belongs to the latter class.

This distinction is remarkably similar to one that Poincare described about 15 years earlier. They were distinguishing between top-down and bottom-up theories. The constructive theory is bottom-up, and the principle theory is top-down.

Einstein was saying that his presentation of relativity was that of a top-down theory. He gave some abstract postulates (indistinguishability of frames, constancy of light speed), and worked out the details as consequences. He was dissatisfied with this presentation, because in early 1908 he complained:

> A physical theory can be satisfactory only if its structures are composed of elementary foundations. The theory of relativity is ultimately just as unsatisfactory as, for example, classical thermodynamics was before Boltzmann interpreted entropy as probability.

Einstein was expressing dissatisfaction with his own top-down description of special relativity. In his 1949 autobiographical notes, Einstein wrote that he had searched for a bottom-up derivation:

> Gradually I despaired of the possibility of discovering the true laws by means of constructive efforts based on known facts. The longer and more desperately I tried, the more I came to the conviction that only the discovery of a universal formal principle could lead us to assured results.

So Einstein was dissatisfied with his 1905 explanation of special relativity, and desperately searched for a more constructive one.

Lorentz's approach was more bottom-up than Einstein's. Lorentz studied the electromagnetic experiments, and then the differential equations for electrodynamics, and then the experiments testing those equations, and then looked for transformations that explained those experiments. The existence of those transformations became Einstein's postulate. His approach was the reverse of Einstein's because he did the detailed theory first, and then abstracted out the abstract principles.

Poincare did special relativity both ways. He wrote technical papers improving on Lorentz's results, and he wrote philosophical papers discussing the high-level principles. He explained the advantage of Lorentz's bottom-up approach:

> Good theories are flexible. Those which have a rigid form and which can not change that form without collapsing really have too little vitality. But if a theory is solid, then it can be cast in diverse forms, it resists all attacks, and its essential meaning remains unaffected. ... Good theories can respond to all objections. Specious arguments have no effect on them, and they also triumph over all serious objections. However, in triumphing they may be transformed.[164]

If you like top-down better than bottom-up, then you may like Einstein's approach better than Lorentz's. But surely Poincare had the superior top-down approach. Among other approaches, Poincare proposed deriving relativity from spacetime geometry. That is the approach that was necessary for general relativity at the time, and it remains the preferred approach today.

There are pedagogic advantages to a top-down approach, and perhaps that partially explains why Lorentz gets so little credit today. But it does not explain why Poincare gets even less credit. Poincare was very much a believer in explaining physics in terms of fundamental principles. He had a list of five or six such principles for physics, and the relativity principle was one of them. He gave multiple derivations of the Lorentz transformations from different principles.

There were even some philosophers who said that Lorentz's theory was unscientific because it was bottom-up. They say that it was *ad hoc* or *phenomenological*. The Latin phrase literally means "for this", and is used to describe a makeshift solution to a specific problem. A more modern term might be a *kluge* or a *hack*. These terms suggest a clumsy fix to an existing program, instead of a redesigned new program. The criticism meant that he devised his formulas as modifications for the specific purpose of explaining the Michelson-Morley and similar experiments, and that his theory might not apply more generally. Einstein was more scientific, they say, because his formulas came from grand principles and theoretical considerations, and not from Michelson-Morley. Poincare made this criticism in 1900, and Lorentz responded with an improved theory in 1904.

At this point, you might think that I am stating this backwards. Isn't science all about reconciling theory with experiment? Yes, it was before Einstein. And it still is to most scientists.

Here is an example of how reasoning might be ad hoc. Suppose a scientist predicts that the weather will be sunny each day next week. But it rains on Tuesday so the scientist revises his prediction to be sunny each day except Tuesday. That would be an ad hoc correction, and it is useless because it doesn't tell you anything about what might happen in subsequent weeks.

Now suppose that particle physics theory assumes that neutrinos are massless. But then an experiment shows that neutrinos have mass, so the theory is modified so that all neutrinos have mass. You might say that the modification is ad hoc, but you might also say that it is measuring a parameter that was previously unknown. Either way it is useful because it successfully predicts other experiments involving neutrinos.

Lorentz and others developed the Lorentz transformation in order to explain Michelson-Morley and other experiments. Poincare argued that it was ad hoc in 1900, and Lorentz responded in 1904:

> POINCARÉ has objected to the existing theory of electric and optical phenomena in moving bodies that, in order to explain MICHELSONS'S negative result, the introduction of a new hypothesis has been required, and that the same necessity may occur each time new facts will be brought to light. Surely, this course of inventing special hypothesis for each new experimental result is somewhat artificial. It would be more satisfactory, if it were possible to show, by means of certain fundamental assumptions, and without neglecting terms of one order of magnitude or another, that many electromagnetic actions are entirely independent of the motion of the system. Some years ago, I have already sought to frame a theory of this kind. I believe now to be able to treat the subject with a better result. The only restriction as regards the velocity will be that it be smaller than that of light.[165]

Poincare's point was that Lorentz's theory explained the experiments for small velocities, but that it would be better to have a relativity principle that held for all velocities. Lorentz's 1904 paper accomplished precisely that. His transformations were essentially the same as he had used in 1899, and also similar to what had been previously used by Voigt and Larmor. Poincare's view was that a theory is not ad hoc if it makes unexpected empirical predictions. Once the Lorentz theory could handle arbitrary velocities, it was no more ad hoc than Einstein's 1905 theory. Lorentz's theory was not limited to Michelson-Morley or to low velocities. It applied to all known electromagnetic situations.

Einstein echoed Poincare's point in a 50-page 1907 relativity review paper, and criticized Lorentz's 1895 theory for being ad hoc.[166] He did not mention Poincare or explain Lorentz's 1904 response. For many years afterwards, many physicists and philosophers have also criticized Lorentz's theory for being ad hoc. They were all complaining about something that had been resolved in 1904.

Einstein inspired a whole generation of theoretical physicists who tried to create top-down theories from postulates and thought experiments, without paying much attention to actual experimental data. Most of this work went into unified field theories, and has been a complete dead end. The progress of physics has been mostly bottom-up, not top-down. Once a new theory is formulated and understood, it can be explained in a top-down manner. That is what happened with relativity, quantum mechanics, and the standard model for particle physics. Perhaps if more physicists understood the history of relativity, they would not try to so hard to replicate something that never happened. The trouble with top-down reasoning is that starting with a bad design leads to a waste of time.

Winning the Nobel Prize

Einstein won a Nobel Prize in Physics, but not for relativity. This fact has long puzzled historians. If Einstein was the greatest physicist, and created the greatest theory, why wouldn't he get a prize for it? All the other great theories of 20th century physics were rewarded with Nobel prizes.

Einstein did get the 1921 prize in 1922, and there is a simple explanation for that. He had become the most famous physicist in the world, by far, and his lack of a prize had become an embarrassment to the Nobel committee. He was being canonized as a secular saint. His prize cited him "for his services to theoretical physics, and especially for his discovery of the law of the photoelectric effect." It was not for discovering that light was quantized, as Planck had already gotten the prize for that. It was for applying Planck's ideas, and with some vague wording that would allow you to believe that he was also being credited for relativity, if you wish. But the Swedish telegram to Einstein said that the prize was not for relativity.[167]

Einstein had been nominated for a Nobel Prize many times previously, and rejected. Poincare was nominated even more times.[168] Sometimes people explain that relativity was too radical to be accepted, or that the committee was waiting for experimental confirmation. But neither of those is true. The relativistic mass of an electron had been measured, and confirmed special relativity formulas. Experiments like Michelson-Morley were done and redone many times. Relativity had been widely accepted for over ten years.

So why wasn't a prize given explicitly for relativity? I think that the most obvious explanation is that the Nobel committee had already given a physics prize for relativity to Lorentz in 1902, and did not want to give another one. The prize cited his work on magnetism and radiation, and not relativity, but his work on the relativity of electromagnetic radiation was part of why he was nominated. A prize for relativity would have surely been shared with Lorentz, but no one else got two physics prizes. It was not until 1972 that someone got a second physics prize.

The other deserving recipient of a special relativity prize would have been Poincare, but he died in 1912 and Nobel prizes are not given posthumously. It seems likely that the reason Einstein did not get a prize for relativity was that he did not invent relativity.

A number of other Nobel prizes were given for physics related to relativity. Besides the 1902 and 1921 prizes, Michelson got the 1907 prize for his light measurement experiments, Paul Dirac got the 1933 prize for a quantum theory of the electron that was invariant under the Poincare group. The 1906 prize went for experimental confirmation of electron theory. Laue and Planck did some important early work in relativity, and they got prizes for their non-relativity works in 1914 and 1918. Pauli got the 1945 prize for his exclusion principle that no two fermions can occupy the same state in a relativistic quantum field theory. The American physicist Richard Feynman and others got the 1965 prize for quantum electrodynamics, the relativistic quantum theory of electromagnetic fields. Prizes in 1978, 1983, and 1993 were given for confirming predictions of general relativity. Those were for work related to the big bang, black holes, and gravity waves, respectively. Many additional prizes were given for work on the standard model of particle physics, a relativistic theory.[169]

A relativity prize to Einstein would have been very controversial. Hardly anyone even mentions relativity without explicitly giving sole credit to Einstein, as in "Einstein's theory of relativity". No other theory in all of science is so directly and personally identified with one man. And yet the bulk of Einstein's fame comes from publishing ideas that were previously published by others. Many physicists of the day did not think that Einstein deserved a prize for relativity.

Einstein did not attend the Nobel medal ceremony, and went to Japan instead where he gave a lecture on "How I Created the Theory of Relativity".[170] He gave the impression that he created relativity entirely himself in a few flashes of brilliance, and minimized the influence of others. He does claim that he was trying to explain Michelson's experiments, but modern historians doubt this and question whether he even knew about them. He gave dozens of interviews throughout his life in which he told similar stories.

The justification for giving Einstein the Nobel Prize was that his work on the photoelectric effect made him a founder of quantum mechanics. In a 1949 book honoring Albert Einstein, the Danish physicist Neils Bohr wrote:

> Einstein's great original contribution to quantum theory (1905) was just the recognition of how physical phenomena like the photo-effect may depend directly on individual quantum effects. With unfailing intuition Einstein thus was led step by step to the conclusion that any radiation process involves the emission or absorption of individual light quanta or "photons" with energy and momentum $E = hf$ and $P = hs$ respectively, where h is Planck's constant, while f and s are the number of vibrations per unit time and the number of waves per unit length, respectively.

Bohr really was a founder of quantum mechanics, so his opinion is worth something. But it was Planck who said in 1900 that light was absorbed and emitted in discrete quanta. After all, the constant h is called *Planck's constant* and not Einstein's constant. Planck got a Nobel Prize for it in 1918, and the German physicist Philipp Lenard got one in 1905 for confirming it with the photo-electric effect. Bohr is really crediting Einstein for recognizing what Planck did.

What Einstein got wrong

Einstein is considered the 20th century's greatest physicist, and his 1905 special relativity paper is considered the century's greatest physics paper. Harvard professor Peter Galison wrote:

> Einstein's 1905 "On the Electrodynamics of Moving Bodies" became the best-known physics paper of the twentieth century. Einstein's argument, as it is usually understood, departs so radically from the older, "practical" world of classical mechanics that the work has become a model of the revolutionary divide. Part philosophy and part physics, this rethinking of distant simultaneity has come to symbolize the irresolvable break of twentieth-century physics from that of the nineteenth.[171]

And yet Einstein's 1905 paper had some serious shortcomings. The most obvious one is that it fails to cite any references to the scientific literature. The failure is extremely odd. The best mathematical physicists in Europe had been writing papers on the subject for ten years, and Einstein did not cite any of them. There is one oblique reference to Lorentz's 1895 result on the subject, but even that reference does not mention Lorentz's name. He uses some of Poincare's terminology without mentioning Poincare's name. Einstein is not just sloppy; he is artfully vague about his sources.

Einstein was a Swiss patent examiner. It is a job that requires being familiar with current publications, and scrupulously citing what is called the *prior art*. Patents are only awarded for inventions that demonstrably surpass what was previously publicly known. A good patent application explains why the invention is novel and useful, and the examiner compares it to the published prior art. Patents can be invalidated if an inventor fails to disclose relevant prior art. It is particularly strange for a patent examiner to avoid references to previous work.

Einstein is alleged to have said, "The secret to creativity is knowing how to hide your sources." That is unconfirmed, but he did say something similar in a 1907 relativity paper:

> It appears to me that it is the nature of the business that what follows has already been partly solved by other authors. Despite that fact, since the issues of concern are here addressed from a new point of view, I am entitled to leave out a thoroughly pedantic survey of the literature …

He is also supposed to have said, "Man usually avoids attributing cleverness to somebody else – unless it is an enemy." He continued to misrepresent his sources in a 1907 relativity survey and in a 1909 survey of radiation theories.[172] He pretended in 1909 that Lorentz's theory was defined by his 1895 book only, and did not mention Poincare at all:

> This contradiction was chiefly eliminated by the pioneering work of H. A. Lorentz in 1895. ... Only one experiment seemed incompatible with Lorentz's theory, namely, the interference experiment of Michelson and Morley. ...
>
> This state of affairs was very unsatisfying. The only useful and fundamentally basic theory was that of Lorentz, which depended on a completely immobile aether. The Earth had to be seen as moving relative to this aether.

Einstein was saying that Lorentz's 1895 theory was able to explain most of the relativity experiments but it used approximations and left open the possibility that a more sensitive experiment could detect the Earth's motion. The 1887 Michelson-Morley experiment was more sensitive, and therefore not fully explained by Lorentz's 1895 theory. Einstein gives the impression that his 1905 relativity paper solved this problem by finding a theory accurate enough to explain Michelson-Morley.

But Einstein ignores the fact that Lorentz and Poincare had improved their theory a lot between 1895 and 1905, and it was completely compatible with the Michelson-Morley experiment. Their theory did not depend on the aether any more than Einstein's. In fact Lorentz and Poincare were largely motivated by explaining that experiment, while Einstein was not. There is no way Einstein could have misunderstood that.

Einstein goes on to try to imply that there is something wrong with Lorentz's theory:

> Superficial consideration suggests that the essential parts of Lorentz's theory cannot be reconciled with the relativity principle.

Then he says that relativity principle can be reconciled if "the hitherto prevailing transformation equations ... are abandoned", and if transformations preserving the metric tensor are used instead. He is saying that the problem is solved by the Lorentz transformations and the metric tensor, without mentioning Lorentz, Poincare, or Minkowski. He concludes that "This path leads to the so-called relativity theory." He is literally correct that "superficial consideration" of Lorentz's 1895 theory might lead one to believe that it was wrong, but extremely deceptive because Lorentz had reconciled the theory, and showed that the superficial consideration was incorrect. Einstein's paper is titled, "The Development of Our Views ...", but it really describes the development of Lorentz's views, and passes them off as his own.

Sometimes Einstein even ignored Lorentz's 1895 work. As late as 1918, Einstein was still criticizing Lorentz for his 1892 derivation of the Fresnel drag coefficient, and ignoring Lorentz's simpler and more relativistic 1895 derivation.[173]

Lorentz's relativity is now called the *Lorentz aether theory*, in order to emphasize the alleged dependence on the immobile aether, and to show its obsolescence. But no one called it that at the time because his theory had very little to do with the aether. It was called *electron theory*. His 1895 book denied the mechanical properties of the aether in Fresnel's theory, and said that he was leaving an understanding of the aether properties for further research. He said that it did not make sense to say that the aether was at absolute rest, but spoke of the aether being at rest anyway, as a figure of speech. He defined the phrase to mean that no part of the aether moved against another part of the aether, as in Fresnel's aether drift theory.[174] Lorentz's 1895 approach was very similar to Einstein's in 1905 -- Lorentz said that he was avoiding speculation about the aether while Einstein said that the aether was superfluous to his derivation.

Einstein goes on in his 1909 survey to pretend that he invented the relativity principle from Michelson's experiment:

> Michelson's experiment suggests the axiom that all phenomena obey the same laws relative to the Earth's reference frame or, more generally, relative to any reference frame in unaccelerated motion. For brevity, let us call this postulate the relativity principle.

Einstein acts as if he deduced the relativity principle from Michelson's experiment, but others did that many years earlier, and it was Poincare who called that postulate the relativity principle. Einstein goes on to describe the invariance of the metric tensor, as Poincare did in 1905, and then says, "This path leads to the so-called relativity theory." But again, there is no mention of Poincare.

Historians have tried to trace what Einstein knew before writing his 1905 paper, and there is no agreement. Some deny that he was influenced by Michelson's experiments. Einstein only admits to reading Lorentz's 1895 book, but it seems certain that he had access to later papers. He was a patent examiner, which is a job involving staying up-to-date on the published literature. He wrote 21 reviews in 1905 for a journal that also reviewed Lorentz's 1904 relativity paper, and that described the Lorentz transformations in the review.[175] He read Poincare's 1902 book that described the principle of relativity. (He was fluent in French, but probably read the 1904 German translation.) The relativity principle and the Lorentz transformations are the core concepts of Einstein's paper.

It is possible that Einstein did not know about the early work on Lorentz transformations by Voigt and Larmor. They wrote brilliant papers on the subject, but their work was not widely appreciated. Voigt discovered transformations similar to the Lorentz transformations as early as 1887, and tried to use them to explain Michelson-Morley. He also coined the term *tensor* in 1899. Larmor also had similar transformations in 1897, and explained them in his 1900 book, *Aether and Matter*. The book had the wordy subtitle, "a development of the dynamical relations of the aether to material systems on the basis of the atomic constitution of matter including a discussion of the influence of the earth's motion on optical phenomena". The discussion of the Earth's motion was a demonstration that Lorentz transformations preserve Maxwell's equations well enough to explain the Michelson-Morley experiment. But Einstein had to know about the work of Lorentz and Poincare, as they were two of the most famous and respected names in Europe and their work was well-known. Even if Einstein did not know about some of their work, he surely found out quickly after 1905. But with rare exceptions, Einstein spent the rest of his life refusing to give any credit to Voigt, Larmor, Lorentz, and Poincare. (One exception was that Einstein wrote a 1906 paper acknowledging Poincare's priority for $E=mc^2$ in a 1900 paper. Einstein sometimes mentioned Lorentz, but usually to denigrate his work.)

Einstein made his share of scientific and mathematical mistakes. According to a recent book, *Einstein's Mistakes: The Human Failings of Genius*, Einstein published seven different derivations of $E=mc^2$, and all of them have mistakes.[176] The 1905 paper has a faulty formula for relativistic mass, even though Lorentz had previously published the correct formula. Planck corrected Einstein in a 1906 paper. Einstein also unnecessarily made low velocity approximations for the Lorentz force law, and for mass-energy equivalence. Planck corrected these also, and Lorentz had his force law right earlier.

Einstein's 1905 paper has deeper problems. The main technical content of the paper was to show how to derive the Lorentz transformations from the Poincare relativity principle and the constancy of the speed of light. But not even that was correct. His argument was sloppy, and scholars do not agree about what he was intending.[177] He makes several hidden assumptions, like space being homogeneous and isotropic, as well as other hidden assumptions.[178] These are fairly minor technical oversights, but it would have been better to spell out the assumptions. We are used to space being isotropic (all directions alike), but to a moving observer, some directions are contracted differently from others.

The essence of special relativity is the symmetry of spacetime. That is, understanding special relativity means understanding how symmetries relate space and time as a single spacetime. Being homogeneous and isotropic are properties of those symmetries, and are obvious from an understanding of the symmetry group.

Einstein did not put space and time together to form spacetime. He did not notice that the Lorentz transformations form a group. He did not use the Lorentz group to gain a better understanding of spacetime. He did not recognize the 4-dimensional non-Euclidean geometry. He attempted to separate the kinematics from Lorentz's theory, but he was not able to demonstrate the covariance of Maxwell's equations for electromagnetism. He just used Lorentz's formulas for transforming the electric and magnetic fields, and failed to show that the field transformations were the natural consequences of the Lorentz group acting on spacetime.

These failures might be excusable, except that Poincare had already done them correctly. Poincare analyzed the group structure of the symmetries, and the metric structure of spacetime. He boldly declared that all of the laws of physics should obey these symmetries. He correctly proved that Maxwell's equations and the Lorentz force law obey the symmetries, improving on results of Lorentz and himself. He proposed a relativistic theory of gravitation.

You might think that understanding spacetime as a 4-dimensional space was obvious from Einstein's 1905 paper, but it wasn't to him or anyone else at the time. Even after Minkowski improved on Poincare's presentation of spacetime a couple of years later, Einstein did not accept it.[179] He said "Since the mathematicians have grabbed hold of the theory of relativity, I myself no longer understand it."

What Minkowski did, more than anything else, was to reformulate relativity with spacetime covariance as the central unifying concept. That meant that he defined a 4-dimensional geometry for spacetime, and required the physical variables to obey the spacetime symmetries. Einstein did not appreciate the covariance concept until many years later.

Even after Poincare introduced 4-dimensional spacetime in 1905 and Minkowski declared space and time inseparable in 1908, Einstein denounced it as "superfluous learnedness". And even after his friend and collaborator Grossmann found relativistic field equations for gravity in 1913, Einstein was still looking for equations that separated space and time. Einstein's proposals violated the relativity principle, and were incompatible with special relativity.

Here is an example to show the importance of the 4-dimensional spacetime view. Einstein endorsed Planck's idea that light was emitted in particles (or quanta) with energy. If the symmetries of space and time apply to all the laws of physics, then the same symmetries must apply to momentum and energy. That is, momentum and energy have to be related in the same way that space and time are related. If Einstein had understood this, then he would have declared light to have momentum as well as energy. He did not until several years later.

The Russian physicist A.A. Logunov wrote a 2005 book on *Henri Poincare and Relativity Theory*. He explains in detail how Poincare's 1905 theory was superior to Einstein's. He also explains how distinguished historians and physicists have misunderstood the contributions of Poincare and Einstein for a century. They have said all sorts of crazy and false things to exaggerate what Einstein did, and minimize or ignore what Poincare did.

Einstein's 1905 paper on special relativity is the most overrated paper ever written. No other paper has been so thoroughly praised, and yet be so dishonestly unoriginal. From it, Einstein learned that he could write a pretentious paper and get away with it.

Einstein's second most famous paper was his 1905 paper on light being transmitted in discrete units, now called photons, and not as waves. He was using Planck's 1900 idea that light was emitted and absorbed as photons, and applying it to explain the photoelectric effect. Using the photon idea to explain the photoelectric effect was a good idea, but rejecting the wave theory of light was not. Our modern understanding of light is closer to Planck's, where light is transmitted as a wave and emitted and absorbed as photons.

Einstein made a number of mistakes on his way to general relativity. His early attempts violated both of his special relativity postulates, and even tried to explain gravity as being caused by changes in the speed of light. The most famous mistake was his "hole argument" which sidetracked him for two years. He was confused about the symmetries generated by coordinate transformations. He was concerned that gravitational field equations could never be solved because the variables look different in different coordinates. But the coordinates can be chosen arbitrarily, and he should not have expected them to be determined by physical arguments. Just as he was confused in 1905 about how electromagnetic fields behave under Lorentz transformations, he was confused in 1913 about how gravitational fields behave under coordinate transformations.

In a 1913 letter to a colleague, Einstein wrote:

> The gravitation affair has been clarified to my complete satisfaction. One can specifically prove that generally covariant equations that completely determine the field from the matter tensor cannot exist.[180]

Einstein was claiming that it was impossible to have relativistic field equations for gravity. He was completely wrong. His friend Grossmann showed him how covariant equations could determine the gravitational field in a 1913 paper, and Einstein didn't get it until Hilbert showed him again in late 1915. Einstein's proof was fallacious.

In July 1915, when Einstein was struggling to understand the superiority of Grossmann's equations, he was also actively sabotaging his friend's reputation. He wrote to Sommerfeld:

> Grossmann will never lay claim to being co-discoverer. He only helped in guiding me through the mathematical literature but contributed nothing of substance to the results.

But it was Grossmann's equations that led to a relativistic theory of gravity, not Einstein's.

The two main experimental tests of general relativity were the apparent deflection of starlight and the precession of Mercury's orbit. Einstein published faulty analyses of both of these before settling on correct ones.

The two most striking cosmological predictions of general relativity are the big bang and black holes. These are both inevitable consequences of the field equations and certain physical assumptions, and are both very well confirmed today. But Einstein missed the boat on both. When others discovered them, Einstein criticized them as incorrect.[181] He only realized his errors many years later.

The *big bang* theory is a consequence of a dynamic universe. If the so-called "fixed stars" are subject to the same law of gravity as our solar system and the rest of the universe, then they will exert attractive forces on each other. The stars will not just stand still any more than the planet Mars can stand still. Levi-Civita and de Sitter published relativistic cosmological models of an expanding universe in 1917. The Russian mathematician Alexander Friedmann discovered an exact solution for an expanding universe in 1922, and sent it personally to Einstein to be published. After it was published, Einstein wrote a letter to the journal falsely claiming that the model did not satisfy the field equations. When the Belgian physicist Georges Lemaitre tried to persuade him of the expansion with some excellent theory and data at a 1927 conference, Einstein said, "Your calculations are correct, but your physical insight is abominable." Einstein only admitted his blunder after everyone else had accepted the expansion of the universe in about 1930.

The German physicist Karl Schwarzschild published a relativistic model of the gravitational field of a point mass in 1916, and it was later recognized as a model of a black hole. The concept of a black hole is much older, as in 1796 Laplace promoted the idea that a star could be so dense that no light could escape its gravity. As late as 1939, Einstein published a paper claiming to offer "a clear understanding as to why these 'Schwarzschild singularities' do not exist in physical reality."[182] Today, thousands of black holes have been found, including a giant one at the center of our Milky Way galaxy.

Einstein often explained that general relativity was based on Mach's principle. Ernst Mach was an Austrian physicist and philosopher who enunciated a number of profound-sounding ideas in the late 1800s. He did not believe in relativity when he died in 1916, or even atoms. There is no agreement on just what his principle is, or whether it is true. After many years of trying to explain it, Einstein wrote in 1954, "As a matter of fact, one should no longer speak of Mach's principle at all".

Einstein published some experimental work in 1915. He had an idea for measuring a certain gyromagnetic ratio and he had a theoretical reason for believing that the ratio would be 1.0. He got Lorentz's son-in-law Johannes de Haas to help him, and they claimed to observe the value 1.02. The trouble was that everyone else who did the experiment got a value close to 2.0. It appeared that Einstein got a value to match his faulty prediction.[183]

The most striking prediction of $E=mc^2$ was the atomic bomb. But Einstein never saw it coming, and his only role was to sign a letter to President Roosevelt once Szilard convinced him of the possibility. He later said that he regretted the letter.

Einstein did not contribute much to quantum mechanics. He was the most famous physicist in the world when the most important new theories were being developed, and he had little to add. Somehow the father of modern physics had nothing say about the essence of modern physics. He said, "I am convinced God does not play dice", and quantum mechanics theory was contrary to his beliefs. He particularly disliked *quantum entanglement*, and said that it was spooky. The theory says that two entangled particles could have the property that a measurement of one could give information about the state of the other. His criticisms became irrelevant, as the theory was wildly successful. He became an embarrassment to his fellow physicists. Pauli once wrote to Heisenberg that, "Einstein has once again expressed himself publicly on quantum mechanics. ... this is a catastrophe every time it happens."

Einstein had an almost religious belief in determinism. To him, causality meant that past events must determine future events, with nothing left to chance. He never really accepted the sufficiency of relativistic causality, which is rooted in Poincare covariance. As his obituary said, "he would not admit the ultimate validity of any theory based on chance of indeterminacy."[184] There are a few physicists today who pursue deterministic theories, but they do not reject quantum field theory and its associated probabilities.

One of Einstein's most famous papers was a 1935 co-authored criticism of quantum uncertainty and entanglement.[185] It argued that quantum mechanics should not be considered a complete theory because there are scenarios in which the theory explains experiments without fully accounting for the underlying physical reality. It described the Heisenberg uncertainty principle, and said, "No reasonable definition of reality could be expected to permit this."

Some physicists thought that they had learned that incomplete theory lesson from Einstein, as his 1905 relativity paper describes observations without attempting to explain physically why measuring rods contract. He believed that *hidden variables* would be found to fully describe reality, and thereby eliminate his philosophical objections. He was proved wrong by a combination of mathematical arguments and physical experiments. The British physicist John S. Bell showed in 1964 that a local hidden variable theory is incompatible with quantum mechanics, and all of the experiments confirmed quantum mechanics. Hidden variable quantum theories have therefore been discredited. The modern view is that multiple interpretations of quantum mechanics are possible, and that there is no consensus about the underlying reality.

Einstein is famous for his work on mathematical theories of physics, but not for any new mathematics. The American mathematical physicist Freeman Dyson said, "Einstein was not a mathematician ... he had no interest in pure mathematics, and he had no technical skill as a mathematician."[186] Hilbert once remarked, referring to the German city where he taught, "Every boy in the streets of Goettingen understands more about four-dimensional geometry than Einstein."[187]

Besides trying to undermine quantum mechanics, Einstein spent his last forty years searching for a unified field theory. The search was a failure. He not only failed to unify anything, his theories did not even address the most important aspects of physics, such as fermions, nuclear forces, and quantum mechanics. Hardly anything he did after 1915 was of much value. The public success of the 1919 eclipse went to his head. Some people credit him with causing others to clarify quantum mechanics in response to his criticisms, but that's about all.

Even though gravity waves were proposed as a consequence of relativity by Poincare as early as 1905, Einstein co-authored a paper in 1936 claiming to prove that gravity waves do not exist. When the referee correctly pointed out that his argument was completely wrong, he wrote the editor a nasty letter and re-submitted the paper elsewhere.[188]

Weyl invented *gauge theory* as a way of combining general relativity and electromagnetism in a brilliant 1918 paper. He had an interpretation of electromagnetism in terms of an overall distance scale factor, or gauge, and suggested that we might not notice a change in gauge. Einstein published an appendix arguing that the interpretation was unphysical. Einstein had a point, but failed to appreciate the significance of the theory. Weyl combined it with quantum mechanics in a 1929 paper. That formulation is used in all quantum field theory today, and is essential to understanding electrons and photons.[189]

Weyl's gauge theory resulted in a geometrical formulation of electromagnetism that completed the work of Maxwell, Poincare, and Minkowski. Einstein completely missed this, and it is not clear that he ever understood it.

At various times Einstein announced a new unified field theory. He announced his first one in 1925, and his 1928 announcement caused so much excitement that the *New York Herald Tribune* newspaper published his paper, complete with Greek-letter formulas. *New York Times* headlines in 1949 and 1952 said that he found a "master key to the universe" and a "new theory to unify law of the cosmos". Seven times he proposed theories in which some variant of the metric tensor was supposed to be the new aether. But he never had anything except poor attempts at a non-quantum unification of gravity with electromagnetism, something that Weyl and others had successfully done. Einstein appeared to not understand Maxwell's equations, and his theories were unable to deal with the simplest properties of electric charges. When his colleagues rejected his work, he complained that "physicists have no understanding of logical and philosophical arguments."[190]

This sort of dramatic publicity would have been justified for the really important breakthroughs in quantum mechanics during the same time period. Born made some of those breakthroughs, and after Dirac incorporated relativity in 1928, he was so excited that he announced, "Physics as we know it will be over in six months." The next year, Dirac said, "The underlying physical laws necessary for the mathematical theory of a large part of physics and the whole of chemistry are thus completely known".[191] The biggest questions in physics were being answered. But none of that made it into the popular newspapers. They were reporting on Einstein's foolish ideas while others were making spectacular progress.

All great scientists have gotten things wrong, and perhaps we should focus on what they did right. But Einstein is credited so much in excess of what he did, that we need to look at his faults to put him in perspective.

Separating spacetime from electromagnetism

Relativity emerged from electromagnetism, and became a theory of space and time. Crediting Einstein for special relativity is usually based on him being the one to separate out the electromagnetism, and formulate relativity as a spacetime theory. But he had nothing to do with it.

Minkowski's 1908 formulation of relativity was as a spacetime theory. He described a geometrical structure on 4-dimensional spacetime, and the simultaneity, length contraction, time dilation, and invariant speed of light are all consequences of that structure. His starting point was Poincare's 1905 work, which also described relativity as a spacetime theory.

The Russian physicist V. L. Ginzburg got a Nobel Prize in 2003 for superconductivity physics. He discusses the origin of the special theory of relativity (STR) in his 2001 autobiography.[192] He is a big Einstein fan, and is very upset that some people credit Poincare and Lorentz for special relativity. He writes:

> it should be emphasized once again that the STR is a theory based precisely on the relativity principle and the Lorentz transformations. Once this basic premise is understood we can discuss the origin of the theory, its authors, and their intentions. … [p.226]

After this, Ginzburg says that he does not understand how Poincare's explanation goes further than Lorentz. Part of the problem is that Poincare credits Lorentz for this basic premise:

> Lorentz's idea may be summed up like this: if we are able to impress a translation upon an entire system without modifying any observable phenomena, it is because the equations of an electromagnetic medium are unaltered by certain transformations, which we will *call Lorentz transformations*. Two systems, one of which is at rest, the other in translation, become thereby exact images of each other.[193]

Where Poincare goes further than Lorentz is by proving that the equations in one system are deducible from the equations in the other system, by applying those transformations of space and time. In a word, Poincare proved covariance, while Lorentz and Einstein did not. Here is Einstein's 1905 version:

> [Examples] suggest rather that, as has already been shown to the first order of small quantities, the same laws of electrodynamics and optics will be valid for all frames of reference for which the equations of mechanics hold good. We will raise this conjecture (the purport of which will hereafter be called the "Principle of Relativity") to the status of a postulate, ...

This refers to Lorentz's 1895 theorem of the corresponding states, which Lorentz proved in a first order approximation, and ignores Lorentz's proof to all orders in 1904. This also seems to refer to Poincare's 1900 conjecture that the theorem holds to all orders. Einstein gives a presentation of the consequences of Lorentz's theorem, but goes no further than what Lorentz and Poincare had already published. The modern notion of relativistic covariance is based on what Poincare proved in 1905, not what Lorentz and Einstein proved. Feynman correctly credits Poincare for this crucial idea:

> It was Poincaré's suggestion to make this analysis of what you can do to the equations and leave them alone. It was Poincaré's attitude to pay attention to the symmetries of physical laws.[194]

The distinction between what Poincare and Einstein proved in 1905 is widely misunderstood. Nearly everyone falsely credits Einstein for proving the Lorentz covariance of Maxwell's equations. Some people say Poincare and Einstein both proved it. But there is a big difference between what Poincare and Einstein did. It is a little tricky to explain, but it goes to the heart of what relativity is all about and how it was discovered. The difference is in how they use the relativity principle.

Einstein mimicked Poincare's terminology for the relativity principle, but meant something different by it. Einstein combined the relativity principle with Maxwell's equations in order to relate electromagnetic variables in different frames. Only Poincare actually proved that Maxwell's equations have the property that applying the spacetime Lorentz transformation gives those same Maxwell's equations in a different frame. Einstein doesn't really prove any such thing, and instead postulates a refinement of Lorentz's 1895 theorem of the corresponding states. In short, Einstein postulated something similar to what Lorentz and Poincare proved.

Put another way, Poincare's relativity principle says that the laws of physics obey a Lorentz group symmetry, while Einstein's says only that the laws of physics take the same form in any frame. To Poincare, the laws of physics are formulated with covariant formulas on spacetime, so that the relativity principle is a consequence of the geometry of spacetime. Thus he proves the covariance of Maxwell's equations, and that every frame has the same electromagnetic properties. Einstein has to assume as a postulate that Maxwell's equations hold in the same way, because he does not have the concept of how to transform the equations from purely geometrical assumptions. As Feynman said, it was Poincare who truly introduced geometrical spacetime symmetry to physical laws.

Put still another way, the Lorentz transformation had different interpretations. For Lorentz and his theorem of the corresponding states, the transformation had to be defined to act on all the physical variables, including the distance, time, electric charge, and magnetic field. To Poincare, the transformation just acts on distance and time, and he deduces the action on all other variables as a consequence of covariance principles. Einstein used Poincare's operationalism to relate the distance and time variables, but he still had to define the transformation on the electromagnetic variables and could not deduce it solely from the spacetime transformation.

Thus Poincare described relativity as a spacetime theory. The electromagnetic relativity was a consequence of the spacetime relativity. Minkowski popularized the idea in 1908, and Einstein adopted it after that. Anyone who compares Poincare to Einstein without mentioning covariance is skipping the heart of electromagnetic relativity.

After misunderstanding Lorentz covariance, Ginzburg has this gushing praise for Einstein:

> That he was the greatest of the great physicists of our century and, perhaps, of all time is of course, important but hardly everything. Einstein always strove for justice, for liberty, and for other human rights, he despised the dark forces, and was a model of noble human dignity. It would be unimaginable for Einstein to start a dispute, let alone a squabble, over priority issues. The same is true for Lorentz and Poincaré. ... Einstein always emphasized the roles played by Poincaré and Lorentz. ... The greatest physicists in the last hundred-odd years – Maxwell, Lorentz, Planck, Einstein, and Bohr – were exceptionally moral persons. A typical characteristic of their morality was aptly expressed by Einstein in one of his mottos, "An honest person must be respected even if he shares opposite views."[195]

No, this is crazy idol worship. Einstein got into several priority disputes, and he never properly credited Lorentz, Poincare, and others. His personal morals and politics were not so great either. Einstein was capable of keeping a secret to protect his image. He had an illegitimate daughter Lieserl in 1902, and his biographers never even found out about her until 1986.

The French professor Michel Paty compares Poincare and Einstein and their contribution to special relativity:

> But their views were very different concerning the theoretical meaning of these results, and only Einstein can be credited of having developped a *theory* of relativity, where the idea of *covariance* is basic and founding. Although the word was coined afterwards, it summarizes indeed the essential of Einstein's 1905 theory (and, so to speak, the 'object' of this theory) : covariance, as the condition put on physical quantities so that the principle of relativity is obeyed, entails the Lorentz formulae of transformation through a redefinition of space and time, and the covariant form of (electro-)dynamical laws. Poincaré also considered covariance, but not as the founding concept. It was entailed from Lorentz formulae of transformation, and these were a consequence of electrodynamical properties as evidenced experi-

> mentally (with a particular emphasis on Michelson-Morley experiment, at variance with Einstein).[196]

This is backwards. Poincare proved covariance in 1905, and Einstein did not. Covariance means transforming spacetime vectors and tensors as a consequence of spacetime coordinate transformations. This fundamentally important concept is absent from Einstein's famous 1905 paper. Einstein does not even have the concept of spacetime, and he certainly does not have the concept of a covariant vector or tensor field on spacetime. He deduces the transformation of electric and magnetic fields by assuming that Maxwell's equations hold in different inertial frames, just as Lorentz did ten years earlier. The main difference between Einstein and Lorentz on this point was that Einstein said that he was using the equations for empty space in the stationary system, while Lorentz said that he was using the equations for the aether. That is, Einstein and Lorentz merely used different terminology for the same thing.

Einstein's 1905 paper describes Maxwell's equations for empty space in two different frames, and says that "electric and magnetic forces do not exist independently of the state of motion of the system of coordinates."[197] This is not covariance. The covariant statement is that the electromagnetic field tensor *does* exist independently of the coordinates. Coordinates are used to do a calculation in a particular frame of reference, but the important physical variables have a covariant meaning that is independent of those coordinates.

Poincare invented 4-vectors on 4-dimensional spacetime in 1905, gave a covariant spacetime formulation of Maxwell's equations, and proved that the equations hold in different inertial frames by applying those spacetime coordinate transformations. You would think that a French scholar would be able to learn that a Frenchman invented relativistic covariance, but physicists have been teaching this wrong for decades.

The British philosopher Harvey R. Brown wrote a book on relativity, and carefully considers the history of special relativity, saying this about Poincare:

> Indeed, the claim that this giant of pure and applied mathematics co-discovered special relativity is not uncommon, and it is not hard to see why. Poincaré was the first to extend the relativity principle to optics and electrodynamics exactly. Whereas

> Lorentz, in his theorem of corresponding states, had from 1899 effectively assumed this extension of the relativity principle up to second-order effects, Poincaré took it to hold for all orders. Poincaré was the first to show that Maxwell's equations with source terms are strictly Lorentz covariant. ... Poincaré was the first to use the generalized relativity principle as a constraint on the form of the coordinate transformations. He recognized that the relativity principle implies that the transformations form a group, and in further appealing to spatial isotropy. ... Poincaré was the first to see the connection between Lorentz's 'local time', and the issue of clock synchrony. ... It is fair to say that Poincaré was the first to understand the relativity of simultaneity, and the conventionality of distant simultaneity. Poincaré anticipated Minkowski's interpretation of the Lorentz transformations as a passive, rigid rotation within a four-dimensional pseudo-Euclidean space-time. He was also aware that the the electromagnetic potentials transform in the manner of what is now called a Minkowski 4-vector. He anticipated the major results of relativistic dynamics (and in particular the relativistic relations between force, momentum and velocity), but not $E=mc^2$ in its full generality.[198]

Nevertheless, Brown criticizes Poincare saying, "it is not clear that he had a full appreciation of the modern operational significance attached to coordinate transformations", and questions whether "Poincaré understood either length contraction or time dilation to be a consequence of the coordinate transformations." *Time dilation* refers to moving clocks ticking at a slower rate. Relativity teaches that time is distorted on a moving clock, relative to an observer's clock. To a first approximation, the moving clock ticks at the same rate and requires an adjustment based on the distance between the two clocks and the relative speed. To a second approximation, the moving clock ticks more slowly, relative to the other clock.

> It is true that Poincare wrote a May 1905 letter to Lorentz questioning the necessity of the time dilation. It had been part of the theory for five years, in publications of Lorentz and Larmor, but it was a second order effect that Poincare had tried to ignore. But the letter was quickly followed by another letter where Poincare said that he had changed his ideas and explained that the time dilation formula allows the transformations to form a mathematical group, and he gave the relativistic velocity addi-

> tion formula.[199] His June 1905 (short) paper included those formulas. He was correct, of course, and the time dilation was essential to his theory.

A mathematician like Poincare would be unlikely to say that a length contraction is caused by a coordinate transformation. That would make it sound as if a mathematical symmetry causes a physical effect. But he certainly explained the contraction with formulas in his 1905 papers, and with words in his popular essays.

Brown's major point is to promote Lorentz's physical interpretation of relativity. He criticizes Poincare for adopting the view that relativity is a property of spacetime, instead of being a consequence of electrodynamics. As explained earlier, and by Brown in his book, both views are tenable, but it was Poincare's spacetime view that became accepted by physicists.

The French physicist Thibault Damour also prefers to credit Einstein because of time dilation and some obscure terminological issues:

> Poincare was the first to understand, in 1900, that Lorentz's "local time" t' was more than simply a useful auxiliary quantity, Poincaré had indeed realized that if observers in motion decided to synchronize their watches by exchanging light signals, with the assumption that the duration of the signal transmission between the two observers is the same in both directions, their watches would show, at least to the first order of approximation, Lorentz's "local time", t'. ...
>
> If a clock, seen at rest, ticks once every second, the same clock will seem to tick once every two seconds, when one observes it moving at this high speed. This new physical effect, generally called time dilation, was never imagined to exist before Einstein. While some of the equations manipulated by Lorentz and Poincaré were identical to those derived (independently) by Einstein, and indeed contained this factor k modifying the second measured by clocks in motion, Lorentz and Poincaré always thought of time in terms of Newton's absolute time. They never suggested, as Einstein did, that a moving clock would tick at a different rate than that of a clock at rest.[200]

So Damour acknowledges that Lorentz, Poincare, and Einstein all used the same formulas for the time transformations, and that Poincare understood in 1900 that the transformed time would indeed be the local time shown by watches, but somehow argues that Poincare did not imagine what Einstein imagined. Damour is wrong. The relativity principle was essential to Poincare's approach, and the Lorentz group was the mathematical means for realizing the symmetries between the different frames. The group property is the mathematical way of saying that every frame is just like every other frame. As Poincare says in his 1905 letter to Lorentz, the time dilation is necessary for this purpose.

Damour claims that Poincare had some sort of flawed understanding of time, based on his occasional use of terms like "true time", such as in his 1904 lecture:

> The watches thus constructed will therefore not show the true time, they will show what might be called local time, with the effect that one of them will run late with respect to the other.[201]

Damour says that this proves that Poincare believed in the obsolete concept of true time. Similar reasoning would indicate that various prominent atheists must believe in God, because they are always arguing that events are not attributable to God. A straightforward reading of Poincare is that he has a theory about local time, not true time. His papers have formulas for local time, but no formulas for true time. He just calls it "time", except when he is crediting Lorentz.

Pauli wrote a relativity book in 1921, and credited Poincare with proving the covariance of Maxwell's equations.[202] For Poincare, the symmetries act on space and time, and the electromagnetic variables are covariant, which means that the transformation of those variables is deducible from the spacetime transformation. Pauli correctly points out that Lorentz's earlier analysis of Maxwell's equations did not truly prove the invariance, because of an assumption that the field intensities are suitably chosen.

Pauli did not seem to realize that Einstein's 1905 paper had the same shortcoming as Lorentz's papers. After crediting Lorentz and Poincare, he wrote:

> As we saw above, Lorentz and Poincaré had taken Maxwell's equations as the basis of their considerations. On the other hand, it is absolutely essential to insist that such a fundamental theorem as the covariance law should be derivable from the

> simplest possible basic assumptions. The credit for having succeeded in doing just this goes to Einstein.

But Einstein did not prove or even state the covariance law in 1905. Pauli claims to follow Einstein in the succeeding pages, but he actually uses Poincare's covariance law and metric tensor. Poincare's formulation is actually simpler, because it is just based on the geometry of space and time.

Writing later in 1956,[203] Pauli described the essence of special relativity as understanding how simultaneity depends on the frame of reference, how the symmetry group leaves invariant a spacetime metric, and how Maxwell's equations are invariant under the group. Pauli credits Poincare and Einstein with independently discovering these ideas, but Poincare published them five years ahead of Einstein.

The Einstein myth

Einstein is not just idolized by Time and Disney; he is also worshiped by experts who should know better. The Washington University (St. Louis) physicist John S. Rigden wrote the book *Einstein 1905: The Standard of Greatness* and explains:

> In this famous June paper, Einstein included no citations. Much of his source material was "in the air" among scientists in 1905, and some of these ideas had been published. Einstein could have cited the work of Lorentz and Poincaré; however, to do so would have been a bit artificial and perhaps even disingenuous. In the development of his special theory of relativity, Einstein did not draw from or build upon the work of others. He adopted two principles as axiomatic, and by means of his intellectual prowess, he brought the unseen consequences of the two principles into full view. At the end of the paper, he thanked his friend, Michele Besso.[204]

Saying that relativity was already "in the air" is a remarkable admission for a book with 180 pages of unrestrained praise for Einstein. He used ideas that were well-known and published, but he did not cite his sources. He may have also used unpublished ideas, as it is known that Poincare gave a lecture at Zurich while Einstein was a student there.[205] Here is some typical praise:

> Although Einstein died in 1955, he remains the standard of greatness. Smart kids are often nicknamed "Einstein." "Hey Einstein," we ask the class genius, "what did you get on the test?" When television commentators want to refer to real intelligence, they mention Einstein. Why Einstein? He was certainly smart, but many people are smart. Einstein, however, is more than simply a symbol of intelligence. When Einstein recognized truths about the natural world by pure acts of mind, he exemplified what is best about being human. And when, through it all, he exuded a noble modesty, he entered the consciousness of all people.[206]

Sometimes the Einstein nickname is too good. A TV commentator once said, "The word 'genius' isn't applicable in football. A genius is a guy like Norman Einstein."[207]

No, Einstein did not recognize truths by pure acts of mind; he recognized them by reading Lorentz, Poincare, and others without crediting them. He did not exude a noble modesty; he dishonestly got famous on the works of others.

This is a strange thing about Einstein. The experts seem to know that special relativity was already "in the air" when Einstein wrote his famous paper, and that he had access to papers on the subject. Nevertheless, the Einstein fans turn his concealment of his sources into praise for his character.

This sort of nonsense is particularly strange coming from a St. Louis professor. Everyone in St. Louis knows about the 1904 St. Louis World's Fair as being the city's finest year. It is famous for popularizing the ice cream cone and other fine American foods. At that fair, Poincare and others lectured about those special relativity ideas that were in the air, and announced an "entirely new mechanics", just a couple of miles from where Prof. Rigden teaches. He could have told a better story by writing a book about what happened in 1904, not 1905.

Here is what he says about that World's Fair:

> These strange ideas were invented to patch over the problems engendered by the ether concept. The patches were offensive, but physicists believed the ether was required in order for light to travel from place to place. In addition to ad hoc remedies, basic ideas were looked at afresh. In 1898, for example, Henri

> Poincaré raised questions about time: "We have no direct intuition about the equality of two time intervals. People who believe they have this intuition are the dupes of an illusion." And in 1904, at the St. Louis World's Fair, Poincaré asked, "What is the aether, how are its molecules arrayed, do they attract or repel each other?" During his remarks about time, Poincaré talked about clock synchronization; Lorentz defined "local time," which Poincaré elaborated further; Poincaré brought the Galilean Newtonian relativity principle into the discussions.[208]

What he is saying is that Lorentz and Poincare had already figured out that clocks slow down in moving frames. They showed that electromagnetism and optics behaved the same way whether you use the aether or not. They showed the relativity of space and time. Their ideas were "offensive" because hardly anyone else beside Poincare believed that clocks would really slow down.

What Einstein did, according to his description, was to take two of the Lorentz-Poincare principles as axiomatic, and then showed how other aspects of the Lorentz-Poincare theory could be deduced. And then not cite Lorentz or Poincare because that would have been a bit artificial and perhaps even disingenuous. By doing that, he exemplified what is best about being human. This is an example of Einstein worship.

Another Washington University physicist, Clifford M. Will, wrote:

> By 1904 Poincaré understood almost everything there was to understand about relativity. In 1904 he journeyed to St. Louis to speak at the scientific congress associated with the World's Fair, on the newly relocated campus of my own institution, Washington University. In reading Poincaré's paper "The Principles of Mathematical Physics", one senses that he is so close to having special relativity that he can almost taste it. Yet he could not take the final leap to the new understanding of time. …
>
> While the great physicists of the day, such as Lorentz, Poincaré and others were struggling to bring all these facts together by proposing concepts such as "internal time", or postulating and then rejecting "aether drift", Einstein's attitude seems to have been similar to that expressed in the American idiom: "if it walks like a duck and quacks like a duck, it's a duck". [209]

The concept was "local time", and it was indeed a crucial breakthrough for special relativity. Will is saying that Einstein recognized that local time is the same as the observed time, but that was also the understanding before Einstein. Local time was invented to explain Michelson-Morley and related experiments, and that explanation only works if local time is the same as the observed time.

Sometimes people credit Einstein because Poincare generously credited Lorentz and Lorentz credited Einstein. In 1927 (after Poincare was dead), Lorentz said:

> Only the true time existed for me. I regarded my transformation of time merely as an heuristic working hypothesis. Thus, the theory of relativity is, in fact, exclusively Einstein's product.

This is an odd admission, because in 1900 Poincare clearly said that local time was the time measured by the local clocks, and attributed the idea to Lorentz. If Lorentz had been more vain, he would have just accepted the credit. Anyone before Einstein would have said that the Lorentz-Poincare relativity theory is that clocks show local time. Even if local time were just a heuristic working hypothesis for Lorentz, it is odd not to credit him anyway. Einstein got his Nobel Prize for a heuristic working hypothesis, and so did a lot of other physicists. For example, the American physicist Murray Gell-Mann got the Nobel Prize in 1969 for discovering quarks, and the official citation praised its heuristic value:

> It has not yet been possible to find individual quarks although they have been eagerly looked for. Gell-Mann's idea is none the less of great heuristic value.[210]

Einstein himself wrote in 1907 that he considered the relativity principle (with the principle of the constancy of the speed of light) "merely as a heuristic principle".[211] He emphasized that the principle only gives an incomplete view of electrodynamics.

Lorentz did not have any conceptual misunderstanding. It appears that what Lorentz was missing was the operational definition of distance and time. While he had the concept of electromagnetic local time in 1895, he did not have a way of consistently defining time in terms of clocks and observers. Poincare published such a method in 1900 and again in his 1904 St. Louis lecture, and Einstein devotes the first couple of sections of his famous 1905 paper to it. Lorentz explains it in his 1906 Columbia University lectures and his 1909 electron theory book. After describing his own electromagnetism theory of moving bodies, he wrote:

> The denominations "effective coordinates", "effective time" etc. of which we have availed ourselves for the sake of facilitating our mode of expression, have prepared us for a very interesting interpretation of the above results, for which we are indebted to Einstein.[212]

He then gives an operational definition of distance and time, without mentioning Poincare. He points out that his presentation of the theory is a little different from Einstein's, and that his might be advantageous if it turns out that that the aether is observable. Lorentz viewed Einstein's work as an interpretation of his own, and that was the view of others at the time. There was just one Lorentz-Einstein theory of relativity.

Lorentz was also missing the concept of a tensor. A tensor automatically comes with a set of transformation rules so that it can be written in different coordinate frames, just like a vector. After he published an improved version of his theorem of the corresponding states in 1904, Poincare wrote him a letter saying that the electric charge density had been transformed incorrectly. It is unlikely that Lorentz understood how to determine the correctness of such a transformation. He was only able to prove a correspondence of a variable in one frame to the other, and not how to deduce one from the other. Einstein was also missing the concept of a tensor, until after Minkowski popularized the use of tensors in relativity several years later, and Grossmann tutored him in tensor analysis.

It may also seem strange to see these brilliant researchers crediting others for their own ideas, but it happens all the time. For example, suppose that a hypothetical Professor Adam writes a brilliant paper with a trivial error in the middle of it. If Professor Bob reads the paper, and agrees that it is essentially correct but spots the error, then he will normally politely and discreetly suggest that Professor Adam correct his minor typo. Likewise, suppose that Professor Bob notices a simple and important deduction that follows easily from the paper, and he is not sure whether Professor Adam was aware of it. A polite and respectful Professor Bob will normally give him the benefit of the doubt, and might write a paper that presumes that Professor Adam was aware of the deduction. Professors want to make their reputations on substantial new work, and not just piggybacking on someone else's papers with trivial observations. So it is not so unusual that Lorentz and Poincare over-credited others. Once Lorentz had proposed the transformations for local time, the idea that clocks would show local is both brilliant and trivial at the same time.

Poincare had no reason to comment on Einstein's 1905 paper. Poincare had already written that Lorentz's relativity paper of 1904 was a "paper of supreme importance."[213] Poincare's 1904 lecture announced a new mechanics for velocities less than light, and Lorentz's paper provided the formulas. Einstein's paper gave an exposition of the new mechanics, but it did not have the sort of advance that Poincare would consider to be of supreme importance.

Encyclopedia Britannica credits Einstein over Poincare with this:

> Henri Poincaré ... to write a paper in 1905 on the motion of the electron. This paper, and others of his at this time, came close to anticipating Albert Einstein's discovery of the theory of special relativity. But Poincaré never took the decisive step of reformulating traditional concepts of space and time into space-time, which was Einstein's most profound achievement. ...
>
> His failure to appreciate Einstein helped to relegate his work in physics to obscurity after the revolutions of special and general relativity.[214]

This is completely backwards. Poincare combined space and time into a 4-dimensional spacetime in his 1905 paper, and Einstein did not. Einstein did not even understand the purpose to combining space and time, namely, covariance. Poincare was a leader in relativity research, and Einstein was a follower who gained fame primarily for writing review papers of the work of others. Poincare's special relativity is what is taught today, while Einstein's method of postulating Lorentz's theorem has been relegated to obscurity.

The British mathematical physicist Roger Penrose credits Lorentz and Poincare for much of relativity before Einstein, and writes:

> Historians still argue about whether or not Poincaré fully appreciated special relativity before Einstein entered the scene. My own point of view would be that whereas this may be true, special relativity was not fully appreciated (either by Poincaré or by Einstein) until Hermann Minkowski presented, in 1908, the four dimensional space time picture. ... Einstein seems not to have

> appreciated the significance of Minkowski's contribution initially, and for about two years he did not take it seriously.

Penrose does not realize that Poincare presented that 4-dimensional spacetime picture in 1905. He credits Poincare in another book by having a chapter on "The special relativity of Einstein and Poincare."[215]

Here is what physicist Banesh Hoffmann wrote about Einstein:

> But the real key to the theory of relativity came to him unexpectedly, after years of bafflement, as he awoke one morning and sat up in bed. Suddenly the pieces of a majestic jigsaw puzzle fell into place with an ease and naturalness that gave him immediate confidence. ... What flashed on Einstein as he sat up in bed that momentous morning was that he would have to give up one of our most cherished notions about time.

No, this did not happen to Einstein in 1905. Lorentz had already published his theory of local time in 1895, and got a Nobel Prize in 1902. Poincare had written a book for the general public in 1902 where he denied absolute time. So that cherished notion of time had already been abandoned by those who understood Lorentz and Poincare.

An Einstein biographer relied on Freeman Dyson to describe the difference between Einstein's and Poincare's special relativity, and Dyson said "Einstein was by temperament revolutionary" and blamed Poincare for trying to relate his theory to previous work by others.[216] Dyson also said about Einstein, "When you put someone on a pedestal, you don't call attention to his feet of clay."[217]

The French physicist Louis de Broglie accused Poincare in 1954 of failing to make the decisive step:

> Somehow Poincaré never made the decisive step and thus let Einstein seize the honor of identifying all the consequences of the relativity principle, in particular, to conduct a profound analysis of the length and time measurement in order to reveal the true physical nature of the relation between time and space that is set by the relativity principle. Why did Poincaré fail to follow through his analysis to the ultimate completion? ... Einstein would have failed without the contributions of Poincaré and Lorentz.

As shown in previous chapters, it was Poincare who identified those spacetime consequences of the relativity principle, and Einstein who failed to take the decisive step.

Hawking's 1988 best-selling book *A Brief History of Time* credits Einstein and Poincare for special relativity:

> Between 1887 and 1905 there were several attempts, most notably by the Dutch physicist Hendrik Lorentz, to explain the result of the Michelson-Morley experiment in terms of objects contracting and clocks slowing down when they moved through the ether. However, in a famous paper in 1905, a hitherto unknown clerk in the Swiss patent office, Albert Einstein, pointed out that the whole idea of an ether was unnecessary, providing one was willing to abandon the idea of absolute time. A similar point was made a few weeks later by a leading French mathematician, Henri Poincaré. Einstein's arguments were closer to physics than those of Poincaré, who regarded this problem as mathematical. Einstein is usually given the credit for the new theory, but Poincare is remembered by having his name attached to an important part of it.
>
> The fundamental postulate of the theory of relativity, as it was called, was that the laws of science should be the same for all freely moving observers, no matter what their speed. ...
>
> If one neglects gravitational effects, as Einstein and Poincaré did in 1905, one has what is called the special theory of relativity.[218]

The book sold nine million copies, far more than any comparable science book. Hawking credits Poincare more than most people, but Poincare had done all those things five years ahead of Einstein.

The Canadian historian (and sociologist) Yves Gingras analyzes the history of ideas by counting citations in published papers. In a 2008 article[219] on why Einstein should be credited for relativity, he points out that the physics community credited Einstein and usually ignored Poincare. He writes:

> And even if the individualistic approach can indeed find good reasons why someone in particular failed to refer to Poincaré, the question remains of the global behaviour of the community; and, short of our accepting a conspiracy theory, this collective behaviour requires a socio-cognitive analysis that transcends particular cases. ...

> Poincaré was never awarded the Nobel Prize for Physics. ... Following Poincaré's death in July 1912, a series of eulogies were published and none of them raised the question of the lack of proper recognition for his contributions to electron theory or suggested that they were equivalent to Einstein's relativity theory, which was, by then, well known in physics.

Gingras acknowledges that a couple of prominent mathematicians recognized Poincare's priority in 1909, but points out that Poincare's papers were not even included in a 1913 collection of basic relativity papers. He argues that the opinion of the physics community should be accepted, regardless of what modern readers might find in the original papers.

Gingras's method may tell us who was popular a century ago, but it won't tell us who discovered special relativity. The content of the papers tells a better story. The core of the theory is the spacetime geometry and electromagnetic covariance. Poincare published these ideas in 1905, and Einstein did not. That 1913 collection included the famous 1908 Minkowski article on spacetime geometry, giving the impression that Minkowski invented it, but Sommerfeld did insert a citation to Poincare. So some people knew that Minkowski and Einstein got the idea from Poincare.

Science is not a popularity contest. Science is about making testable hypotheses and verifiable statements about the natural world. Real scientists expect to be judged on the merits of their works, and not on citation counts. If those physicists actually gave some explanation as to why they preferred Einstein's theory to the Lorentz-Poincare theory, assuming that they did have such a preference, then we could examine the merits of that opinion. But there is no one who recognizes what Poincare published, and still credits Einstein for having better ideas.

No conspiracy theory is needed to explain why physicists did not cite Poincare more. His ideas were bold, startling, and profound. It is remarkable that they were accepted as quickly as they were. It often happens that new and difficult ideas are not understood and accepted right away. It is plausible that German physicists would rather cite a simplified German-language exposition of the theory. At any rate, Poincare's relativity principle, spacetime geometry, and electromagnetic covariance became bedrock principles of 20th century physics.

The historian Olivier Darrigol traced the history of special relativity carefully, and also found that Poincare and Einstein published equivalent theories with the same observational predictions. He finds differences that are mainly verbal. He thinks that it is odd that Poincare used words like "hypothesis" and "apparent", and that he used a synchronization argument that illustrated the necessity of the Lorentz contraction. Darrigol suggests that Einstein would have had some conceptual difficulty with the argument. He also says that Einstein could have gotten the theory from Poincare, but prefers not to speculate:

> On several points – namely, the relativity principle, the physical interpretation of Lorentz's transformations (to first order), and the radiation paradoxes – Poincaré's relevant publications antedated Einstein's relativity paper of 1905 by at least five years, and his suggestions were radically new when they first appeared. On the remaining points, publication was nearly simultaneous.[220]

The nearly simultaneous publication was their 1905 papers. Einstein's famous paper was received on June 30 and published in the leading German physics journal in Sept. 1905. Poincare's paper was published in two parts and in French. The 5-page summary was delivered at the meeting of the Academy of Sciences in Paris on June 5, 1905, and immediately distributed to the major European libraries. The detailed 48-page second part was received in final form on July 23, printed on Dec. 14, and published in Jan. 1906 in the widest-circulation international mathematics journal.[221] Einstein's short mass-energy paper was received on September 27, 1905 and published on Nov. 21.

In fact there is not so much overlap between Einstein's and Poincare's 1905 papers. They both give derivations of Lorentz transformations and the velocity addition law, but Poincare discussed the Lorentz group, the spacetime metric, the covariance of Maxwell's equations, the relativistic action, and gravitation. Einstein only has concepts that Lorentz and Poincare published years earlier, and presented them somewhat differently. My guess is that Einstein saw Poincare's June 1905 paper two weeks before submitting his own, which was already mostly written. Einstein used the Poincare paper to clarify the theory, and maybe got a few ideas. For example, Poincare said that the Lorentz transformations form a group, a concept that Einstein may not have understood. Einstein's paper does say that parallel transformations form a group because of the velocity addition law. Einstein's odd phrasing makes it unlikely that he understood what the group was.

Darrigol ends saying that it is impossible to determine whether the similarities between Poincaré's and Einstein's theories of relativity can be best explained by common circumstances or by direct borrowing.

> In sum, then, Einstein could have borrowed the relativity principle, the definition of simultaneity, the physical interpretation of the Lorentz transformations, and the radiation paradoxes from Poincaré. ... The wisest attitude might be to leave the coincidence of Poincaré's and Einstein's breakthroughs unexplained, ...

Darrigol says that Poincare's ideas were radical and original, and discusses how Einstein might have learned them directly or indirectly. He writes:

> The main problem with this speculation is that Poincaré's name does not appear in Einstein's relativity paper and that Einstein never admitted any such influence in this regard.
>
> One can imagine many reasons for his silence. First, and least plausible, is the possibility that the ambitious Einstein deliberately occulted Poincaré's role in order to get full credit for the new theory. This hardly fits what we know of Einstein's personality.[222]

Actually, it does fit with Einstein's personality. Einstein was a vain egomaniac who repeatedly schemed to get more credit for himself than he deserved, and to avoid crediting others. His famous 1905 special relativity didn't just fail to credit Lorentz and Poincare for their previous work on the subject, it didn't have any references at all! We now know from publication of Einstein's letters that he failed to credit his first wife for help with special relativity, and refused to credit many others. His first wife was a physicist who collaborated with him on relativity. Later papers also frequently failed to credit his sources, and yet he wrote complaint letters when he did not get the credit that he wanted. He used the news media to promote himself more than any other scientist of the day. For the rest of his life he continued to ignore his sources and the contributions of others.

These historians do not want to accuse the great Einstein of misrepresenting about his sources, but there is no other reasonable conclusion.

Einstein was certainly influenced by Poincare, and would have learned all of what Poincare did unless he was willfully avoiding it or did not understand it. Even if Einstein independently reinvented in 1905 some of what Lorentz and Poincare did years earlier, there can be no honest excuse for Einstein refusing to acknowledge Poincare's contributions in later years.[223] Einstein's work was not even as good as what had already been published, and yet we treat him as the greatest genius of all time.
not even as good as what had already been published, and yet we treat him as the greatest genius of all time.

Ad hoc hypotheses

Holton is a German-American physicist, historian, and defender of Einstein's reputation. He responded to Whittaker's book by publishing a 1960 attack on Lorentz's 1904 paper for having *ad hoc hypotheses*. He argued that Einstein should get the credit because of a simpler presentation. Yes, Einstein's 1905 paper was simpler to read, principally because it was more self-contained and because the first half can be read without understanding Maxwell's equations. Relativity textbooks were soon written with even simpler and better explanations.

Holton argues that some hypotheses are ad hoc and some are not, and that notion is at the heart of why he prefers to credit Einstein and not Michelson, FitzGerald, Lorentz, or Poincare. He wrote that the FitzGerald-Lorentz length contraction "has traditionally been called the very paradigm of an ad hoc hypothesis." His clue that it is ad hoc is in the language used to describe it, such as what a professor said in 1892:

> Professor FitzGerald has suggested *a way out of the difficulty* by supposing the size of bodies to be a function of their velocity through the ether.[224]

Apparently a hypothesis is considered ad hoc if it is the consequence of an experiment, rather than a postulate.

Lorentz has been savaged for being ad hoc more than any scientist since Ptolemy. Lorentz discussed Michelson-Morley and other experiments that failed to detect the motion of the Earth. He explained that his 1895 theory was inadequate to fully explain those experiments. He discussed physical properties of electrons. He was finding a theory that explains the observations, and that is a good thing in science, not a bad thing. If he was ad hoc, then most of the great scientific discoveries in history were also ad hoc.

If the ideal is to have simple hypotheses instead of experiment-based (ad hoc) hypotheses, then surely Poincare had the simplest. He gave several approaches, but one of them can be summarized in just two words – imaginary time. He needed just a paragraph to deduce "that the Lorentz transformation is merely a rotation" in 4-dimensional space, if the fourth dimension is imaginary time.[225] Einstein had to assume that Maxwell's equations held the same form in different frames, and none of Poincare's approaches had such a complex assumption.

The Frenchman Roger Cerf recently published an article[226] crediting Einstein over Poincare, relying on the opinions of de Broglie and others. He said that "Poincaré's thinking stopped short of the crucial step" because he once referred to the Lorentz contraction as a "hypothesis", instead of a deduction from Einstein's postulates. But there is nothing wrong with this term, and Poincare wrote a whole book titled *Science and Hypothesis* explaining it. A hypothesis is a scientific statement, possibly generalized from observations, which is subject to verification or falsification by experiment. A hypothesis could also be a useful convention that is not directly testable. Poincare himself gave several ways of deriving the Lorentz transformation, and would have strongly disagreed with the suggestion that any one particular derivation must be preferred over others.

Here is how Poincare defends the use of the word "hypothesis" in the preface to his 1902 book:

> Instead of a summary condemnation we should examine with the utmost care the rôle of hypothesis; we shall then recognise not only that it is necessary, but that in most cases it is legitimate. We shall also see that there are several kinds of hypotheses; that some are verifiable, and when once confirmed by experiment become truths of great fertility; that others may be useful to us in fixing our ideas; and finally, that others are hypotheses only in appearance, and reduce to definitions or to conventions in disguise.

As you can see, Poincare described his ideas with a mathematical precision. It is absurd to claim that Poincare's understanding of the Lorentz contraction was somehow deficient because he once referred to it as a hypothesis. Most of the great scientific advances were formulated as hypotheses. Einstein's famous $E=mc^2$ paper was titled, "Does the

Inertia of a Body Depend Upon Its Energy Content?". He phrased it as a question because he was formulating a hypothesis to be tested.

The commonly recited history of special relativity is so ridiculous that it is a wonder that anyone believes it. One typical textbook, by American physics professor Claude Kacser, gives several pages on the subject, and discusses ad hoc theories, Lorentz, Poincare, Einstein, and Minkowski. About Lorentz, it says:

> The most satisfactory such theory was that developed by H. A. Lorentz between the years 1895 and 1904, since it was not ad hoc. Rather, this theory was based on Maxwell's equations, ... In its experimental consequences the theory of Lorentz as finally developed makes exactly the same predictions as the theory due to Einstein. However the physical models of the universe underlying the two theories are completely opposed.[227]

But the two models were not opposed. The book even includes a translation of the first (kinematic) half of Einstein's 1905 paper, and the reader can see for himself that Einstein did not contradict Lorentz, and that there is no physical model of why the rods contract, nor is there any rejection of any physical model. Neither Lorentz, nor Einstein, nor anyone else at the time saw any inconsistency between Lorentz's and Einstein's versions of the theory. Planck, for example, said only that Einstein's paper was "more general" than Lorentz's.[228] Einstein's view was only seen as something different after he adopted the Poincare-Minkowski geometric view several years later.

Kacser first derives the length contraction from Michelson-Morley, which he states as:

> The velocity of light relative to any observer when measured by that observer is always the same, independent of the state of motion of one observer relative to another.[229]

Note how this statement is very similar to Einstein's postulates. However, Kacser says that FitzGerald's "beautiful explanation" was ad hoc, and "can neither be tested nor refuted." In other words, it was unscientific. The book also says (falsely) that FitzGerald first postulated the contraction as a mechanical friction effect caused by dragging through the aether. The book then gives a derivation of the Lorentz transformation from Einstein's postulates, and the argument is essentially the same as what was given for the FitzGerald contraction from Michelson-Morley, along with the analogous argument for time.

Somehow the reader is supposed to believe that an argument that was unscientific in 1889 suddenly became scientific when it was restated by Einstein in 1905.

Kacser concedes that Poincare had a fully relativistic theory, but denies that he had a "complete theory" like Lorentz and Einstein. Poincare had the same equations and experimental consequences as Lorentz and Einstein, so it is not clear why anyone would think that Einstein's theory was more complete.

It is impossible to read a textbook like Kacser's and understand why Einstein is credited with relativity. Einstein's theory was just as ad hoc as the earlier ad hoc theories, and just as dependent on electromagnetism and the aether as Lorentz's theory. Kacser says that Lorentz uses electromagnetic theory to explain how the length contraction could be a real physical contraction, which ought to be an advantage over Einstein's theory, because Einstein gives no explanation. Kacser also argues about how an "omniscient being" with "nonmaterial meter sticks" might detect something different in Lorentz's theory,[230] but of course Lorentz himself said no such thing, and neither did Einstein or anyone else. Even if Einstein's description had some aesthetic or terminological advantage, it is hard to see why this later theory with the same assumptions, formulas, and physical consequences should be considered so worthy of getting all the credit. A superior theory is not given until about halfway through the book where it starts to explain the geometry of spacetime, which Kacser attributes to Minkowski in 1908.

Stephen G. Brush wrote an article on why relativity was accepted, and noted that many physicists treated Einstein's 1905 paper as a "comparatively minor philosophical gloss" on the previous theory of Lorentz and others. He explains that Einstein's theory was later supposed to be better because it was less ad hoc:

> The preference for novel predictions is often associated or confused with the dislike of ad hoc hypotheses. For example, G. F. FitzGerald explained the negative result of the Michelson-Morley experiment by postulating that "the length of material bodies changes, according as they are moving through the ether or across it," by an amount just sufficient to cancel the expected differences in the times for the light beams to travel the paths along and perpendicular to the earth's motion.

> A similar assumption was later made by H. A. Lorentz as part of his electron theory. The FitzGerald-Lorentz contraction (FLC) was considered ad hoc by physicists because it was not derived from a plausible theory. It is considered ad hoc by philosophers of science because it is not independently testable by any experiment other than the one it was invented to explain. Thus many physicists considered that Einstein's theory was preferable to Lorentz's because it explained the FLC by deriving it from general postulates.[231]

No, this is incorrect. The contraction was considered ad hoc because, as of 1900, Lorentz and Larmor had only shown that it explains the first-order aether drift experiments, and not Michelson-Morley. Lorentz did not demonstrate the higher order invariance until 1904. Larmor is said to have discovered it also, but did not publish it. By 1905, the higher order invariance was not an issue, and Einstein's 1905 paper had no such advantage over the previous work.

Brush says that Einstein's theory was preferable to Lorentz's because of the use of postulates instead of an ad hoc Michelson-Morley. This idea is commonly stated by physicists and philosophers. And it is nonsense. No one ever said anything so ridiculous around 1905. This idea was only cooked up many years later in order try to find some explanation for Einstein having done something better than the previous work. Einstein's theory was just as ad hoc as Lorentz's.

FitzGerald, Lorentz, and Einstein all deduced the length contraction in the same way — as a logical consequence of the speed of light being constant for all observers. Einstein had no plausible theory or explanation for the contraction other than what others had already published. His 1905 paper was the most ad hoc of all of the relativity papers, because he gave no explanation for the contraction, or even any comment on the necessity of an explanation. FitzGerald and Lorentz suggested the electromagnetic explanation, and Poincare gave the spacetime geometry explanation. Einstein just proposed that measuring rods and clocks behave in a way consistent with how Lorentz and the others had interpreted the Michelson-Morley experiment. Einstein also had the most complicated hypotheses, because he had to make assumptions about the validity of Maxwell's equations under Lorentz transformations.

It is also incorrect to say that there was no independent test of Lorentz's theory. His 1899 prediction of electron mass increase with velocity was already being successfully tested in 1901, long before Einstein first said anything on the subject in 1905.

Credit for relativity

The theory of relativity can be divided into the general theory and the special theory, and the special theory can be divided into what could be called the easy part and the hard part. The general theory is harder than even the hard of special relativity, and is discussed in the next chapter. It is the special theory that is considered to have revolutionized 20th century physics, and which is studied so carefully by historians, philosophers, and others.

The easy part of special relativity consists largely of accepting the logical consequences of the Michelson-Morley experiment. Michelson created a sensation when he showed in 1881 that the speed of light appeared to be the same in all directions. Doubts about his experiment caused him to do it much more precisely with his colleague Morley in 1887, with the same results. The prevailing ideas about the symmetries of the universe had to be fundamentally wrong. Either the experiment was wrong, or the Earth was stationary in the aether, or everyone had some misunderstanding about light or measurement.

Fitzgerald proposed a length contraction in 1889 as a consequence of the speed of light being the same for all observers, and so did Lorentz in 1892. By 1900, Lorentz, Larmor, and Poincare had given the corresponding time transformations. Voigt had them also, in a limited context. This was the easy part of the theory. Light was known to be electromagnetic, and Lorentz had proved a theorem in 1895 on how these transformations are compatible with Maxwell's equations for electricity and magnetism.

Maxwell's earlier contributions are also overlooked. He created the first fully relativistic theory. He coined the term "relativity" and wrote about the impossibility of determining absolute velocity. His equations implied a constant speed of light. He suggested an experiment like Michelson's so that it would either detect the Earth's motion or confirm the relativity principle. He was missing the Lorentz transformations, but that would have been the next step.

Einstein's famous 1905 paper is half kinematic and half electromagnetic, and it was an exposition of this easy part of the theory. His kinematic part explains how the Lorentz transformations can be deduced from the speed of light appearing the same to all observers, without mentioning Michelson-Morley, FitzGerald, Lorentz, Poincare, or the others who did related work. His electromagnetic part is an account of Lorentz's 1895 theorem. The paper partially treated the higher-order velocity effects, which was an advantage over the earlier papers of Lorentz and Poincare, but not over their later papers. It was considered an embellishment of Lorentz's theory by physicists at the time. In 1906, Lorentz described Einstein's work as adding an interpretation of local time by means of clock synchronization. Einstein only started trying to distinguish his theory from Lorentz's in 1907, when his main claim was that his approach was less ad hoc because he interpreted Lorentz's "local time" (ortszeit) as time. Minkowski similarly argued in 1908 that Einstein's contribution was to identify Lorentz's local time with the time of an electron.

The hard part of special relativity was published in 1905, but not by Einstein. While Einstein was writing his famous paper, Poincare wrote a 1905 pair of papers with a much deeper and more original analysis, including the geometry of spacetime and the covariance of Maxwell's equations. These ideas are the essence of what is now known as special relativity, but they were not in Einstein's paper and he did not even understand them until several years later. While Poincare's role is not so well-known, the special relativity theory that was so influential in 20th century physics was Poincare's theory, not Einstein's.

This is not to say that the easy part of special relativity was obvious. The constancy of the speed of light, the length contraction, and local time were truly profound ideas that many scientists at the time had trouble accepting. Michelson did not accept them, even though he was the one who did the most persuasive experiment. But they had been published five or more years before Einstein came along, and promoted by Lorentz and Poincare, who were two of the leading theoretical physicists in Europe.

The papers by Lorentz and Poincare were not obscure or overlooked. They were cited in relativity papers by Planck in 1906 and Minkowski in 1907. And yet Einstein neglected Poincare in his relativity review papers, and in dozens of interviews he gave throughout his life. He was never able to explain how his theory was any better than what Poincare had published years earlier. It is not known whether Einstein understood Poincare's 1905 papers, as he always avoided discussing them. The big ideas of those papers, namely the spacetime geometry, electromagnetic covariance, and gravity waves, were not even mentioned in Einstein's 1907 relativity review paper.

Einstein was still working as a Swiss patent clerk in 1908 when Minkowski popularized the hard part of relativity, with the spacetime geometry and covariance. Only then did the physics world get excited about relativity. Publications on relativity skyrocketed, and so did Einstein's career. Then next year he was a physics professor at the University of Zurich, and Minkowski had died of appendicitis. By 1914, Einstein had a high-status professorship at the University of Berlin. Somehow Einstein had gotten credit for what Minkowski published.

The Chilean philosopher Roberto Torretti wrote a 1983 book on relativity that credits Whittaker for illuminating the distinction between the contributions of Lorentz-Einstein and Poincare. It says that they were observationally indistinguishable and mathematically equivalent, and that Poincare's 1905 paper had many important ideas (spacetime, symmetry group, metric, imaginary time, 4-vectors) that Einstein did not have. But Torretti blames Poincare for crediting Lorentz:

> This preamble suggests that Poincaré in any case embraced the main tenets of Lorentz's natural philosophy; namely, that the aether exists and is the seat of the fields of force ... the text ... bears no trace of Poincaré's having ever countenanced a revision of the fundamental concepts of classical kinematics. Poincaré's electrodynamics of moving bodies definitely does not rest, like Einstein's on a modification of the notions of space and time. For this reason, it does not attain to Special Relativity's universal scope.[232]

But all of this is false. Poincare explicitly rejected Lorentz's philosophy, announced a new mechanics, based it all on a new notion of spacetime, and applied it more universally than Einstein, including gravity. That preamble of Poincare in 1905 starts by saying that experiments have repeatedly failed to detect any motion with respect to the aether, and that he was adopting the Postulate of Relativity to say that the impossibility was a general law of nature. His theory does not depend on an aether; it depends on the failure to detect the aether. While he does credit Lorentz, Poincare adopts a different point of view, saying that the difference is analogous to the difference between Ptolemy and Copernicus, and that relativity is "something due to our methods of measurement." Torretti's criticism of Poincare is like attacking Copernicus for embracing the main tenets of Ptolemy's geocentrism. If the Copernicus analogy means anything, then it means that Poincare was rejecting Lorentz's philosophy and proposing a fundamental modification of the notions of space and time.

In contrast, Einstein's famous 1905 paper has no similar rejection of Lorentz's philosophy. The paper does claim to develop a view of kinematics and electromagnetism, but does not claim that any of it is contrary to Lorentz. Towards the end, it says that it is in agreement with "the electrodynamic foundation of Lorentz's theory of the electrodynamics of moving bodies"[233]. Lorentz regarded the paper as an interpretation of his theory, and that is how he described Einstein's work in 1906 lectures on relativity at Columbia University.[234] Follow-up papers by Planck and others did not make any significant distinction between Lorentz's and Einstein's theories.[235] Einstein's own 1906 relativity paper referred to the "theory of Lorentz and Einstein". Cunningham credited relativity to "Lorentz and Einstein" in 1909.[236]

Torretti also wonders "why Poincaré did not greet Einstein's paper as the splendid achievement that it was, a decisive breakthrough in the articulation of insights that Poincaré himself had been the first to suggest."[237] Maybe Poincare was not so impressed. No one was so impressed with Einstein until after Minkowski popularized Poincare's spacetime geometry. For example, the American physical chemist Richard C. Tolman wrote in 1909 that Lorentz offered the only satisfactory explanation of the Michelson-Morley experiment, and that Einstein's additional step was to assert that other such attempts to detect the motion of the Earth will fail.[238] Later on, many scientists joined the Einstein idolizers. For example, Tolman wrote in 1919 that Einstein boldly and completely revolutionized our ideas of space and time.[239]

Part of the confusion in assigning credit is that Poincare generously credits Lorentz for two key concepts — that moving clocks show local time and that the relativity principle holds for all velocities. Poincare's explanation of these concepts goes beyond what Lorentz actually said in his early papers, and it is not known whether other readers made these inferences. Poincare also used the term "Lorentz group", even though Lorentz appears to not have realized that his transformations formed a group. Lorentz does not seem to have even commented on whether he agreed with Poincare or not on these crucial concepts. This is strange because Poincare corresponded privately with Lorentz, and sometimes criticized him publicly. Poincare even published his 1900 relativity paper in a volume dedicated to Lorentz. But after Poincare's death, Lorentz sometimes credited Einstein with these concepts, and occasionally refused to credit Poincare. Lorentz even talked about "Einstein's Principle of Relativity" in 1909,[240] and generously credited Einstein for making him see that Michelson-Morley was a manifestation of that principle. Lorentz seemed to have forgotten that his own 1904 relativity paper was written in response to Poincare making the same point in 1900. The upshot is that it is easy to blame Poincare for the alleged misunderstandings that Lorentz modestly admitted. That is, if Poincare credited Lorentz and Lorentz credited Einstein, then it would appear that Einstein deserves the credit for these concepts. But Poincare published them five years before Einstein, and Poincare's theory never depended on an aether, or on a privileged frame, or on absolute or true time, or on any of those things for which people blame Lorentz.

While Lorentz did not appreciate Poincare's papers at first, he did write a 1914 paper (published in 1921) crediting Poincare over Einstein.[241] Lorentz summarized the history of special relativity, and paid homage to Poincare. He explained how Poincare formulated the postulate of relativity, deduced the "group of relativity" (Lorentz group), and obtained a perfect invariance of the equations of electrodynamics. Lorentz also credited Poincare with showing the necessity of Planck's 1900 quantum hypothesis that light is composed of photons. Lorentz quoted Poincare as saying that the hypothesis "would be, undoubtedly, the greatest and most profound revolution that natural philosophy suffered since Newton". A few years later, the new theory of quantum mechanics proved to be exactly that.

Einstein was a patent examiner, and he understood better than anyone that the patent system only credits inventors for what they specifically claim to have invented. So he claimed to have invented relativity, but he could never explain how his theory was any better than Lorentz's or Poincare's. Those who credit Einstein almost invariably point to some aspect of relativity that he did not even claim himself.

Relativity historian Arthur I. Miller wrote in 1994 that Poincare and Einstein discovered the same theory, with the main difference being that Poincare gave due credit to prior work:

> What Albert Einstein and Henri Poincaré accomplished in 1905 continues to fascinate historians and philosophers of science. Everyone agrees that Einstein and Poincaré confronted the same empirical data for which they formulated identical mathematical formalisms. Most scholars agree that whereas Einstein interpreted the mathematics as a theory of relativity, Poincaré considered it as an improved version of H. A. Lorentz's theory of the electron. Others contend that both men arrived at the special theory of relativity and, consequently, Poincare ought to share the accolades with Einstein. ...
>
> Although it turns out that the affect of Poincaré on Einstein might have been substantial, the honors for special relativity go to Einstein, alone.[242]

So maybe Einstein stole the theory, but he gets all the credit anyway!

Miller gives an assortment of strange reasons for not crediting Poincare. Miller says, "Although worded similarly, the principles of relativity of Poincaré and Einstein differed in content and intent." Miller says that Einstein never wrote to Poincare, or expressed gratitude to Poincare, or engaged in debate with anyone who claimed that Poincare discovered Special Relativity. Meanwhile, Poincare ignored Einstein. Miller says Poincare's 1900 relativity paper had much of the theory, but it had certain technical shortcomings that were not corrected until his 1905 papers. Poincare sought a more ambitious theory, including gravity, while Einstein was concerned with the electrodynamics of an electron. Miller complains that Poincare gave a 1912 lecture on "Relations Between Matter and Ether", and that the aether was so superfluous to his theory that it was not even mentioned. Some of these points seem to undermine his main argument, and suggest that Poincare's 1905 theory was superior to Einstein's.

There are simple explanations for all of this. Einstein did not want to discuss Poincare because there was nothing that he could say without diminishing his own reputation. Maybe he did not even understand Poincare's 1905 papers. Poincare was a world-famous mathematical physicist, and was in no position to complain about not being credited, even if he cared.

Miller argues that Einstein appears to have gotten much of the theory from Poincare, but stops short of accusing Einstein of plagiarism:

> Poincaré's La Science et l'Hypothese which Einstein read in 1904: and Poincare's 1900 essay "La Theorie de la Reaction et la Theorie de Lorentz", which Einstein cited in 1906, could have influenced Einstein's thoughts on simultaneity and the characteristics of light pulses. …
>
> Toward Einstein's realizing the relativity of simultaneity, might he have found useful a particularly pregnant comment by Poincaré in La Science et l'Hypothese to the effect that: "Not only have we no direct intuition of the equality of two durations, but we have not even direct intuition of the simultaneity of two events occurring in two different places. This is what I have explained in an article entitled 'Measurement of Time'." …
>
> The concept of an "event" is central to Einstein's analysis of time and simultaneity in 1905. In its larger sense, Einstein's event is similar to Poincaré's, namely, a phenomenon occurring at a point in space and time measured relative to a reference system. Might Poincaré's passage quoted above from La Science et l'Hypothese be the source for Einstein's use of the term "event," and for his focusing on the distant simultaneity of two events? The similarity between Poincaré's and Einstein's conclusions on how to distinguish between local and distant simultaneity is astonishing.
>
> … the important topic of what Einstein might have learned from Poincaré's papers, particularly concerning the notion of events, distant simultaneity, the importance of attributing a physical interpretation to the local time, the structure of science, frontier issues in physics, and the physics of light pulses emitted from moving sources. This information may well have been significant to Einstein's formulation of the special theory of relativity.[243]

Cosmologist Tony Rothman wrote that Poincare first published the ideas lying at the heart of relativity, and that Einstein was surely profoundly affected. The only reasons he gives for crediting Einstein are that "physicists are notorious for taking history on faith", and that Poincare did not fully explain the speed of light being constant:

> Mathematically, he has more than Einstein does. … Poincaré's paper, alas, is that of a mathematician. Right at the start he sets the speed of light equal to a constant, "for convenience." The

> second, and revolutionary, postulate at the basis of Einstein's relativity is in fact that the speed of light is always observed to be the same constant, regardless of the speed of the observer. Perhaps if Poincaré had been less a brilliant mathematician and more a dumb physicist he would have seen that the whole edifice stands or falls on this "convenience." He didn't.[244]

Poincare chooses units so that the speed of light is equal to one in Maxwell's equation, for convenience. It was not new to use a constant speed of light in those equations, as they had always been written that way. But it was Einstein, not Poincare, who wrote papers several years later proposing that the speed of light was *not* constant, and that gravity could be explained by variation in that speed. So Einstein did not believe that the whole edifice would fall.

The closest Einstein comes to saying that special relativity is a consequence of spacetime properties is this 1905 conclusion:

> we have the proof that, on the basis of our kinematical principles, the electrodynamic foundation of Lorentz's theory of the electrodynamics of moving bodies is in agreement with the principle of relativity.[245]

But he is only showing that electromagnetic relativity is consistent with the kinematics, not that it is a consequence of the kinematics. He does not show that Maxwell's equations are invariant under the Lorentz transformations, and does not claim to show it. He is postulating it. In an earlier section, he interprets the relativity principle as saying that Maxwell's equations are invariant:

> Now the principle of relativity requires that if the Maxwell-Hertz equations for empty space hold good in system K, they also hold good in system k;[246]

This is in contrast to the Poincare-Minkowski approach, which was to make assumptions about the geometry of spacetime, and then to deduce the invariance of Maxwell's equations. That is, they *prove* that if the equations hold in one system, then they hold in another. Einstein has to make his kinematic assumptions as well as his electromagnetic assumptions. He had no real explanation for why the same Lorentz transformations should apply to kinematics as well as to electromagnetic fields. Minkowski said in 1908 that the Lorentz-Einstein usage of the relativity principle was feeble[247] and that he preferred the term *world postulate* for the idea that relativity is a 4-dimensional spacetime phenomenon.[248] Einstein's concept of Lorentz invariance was essentially the same as Lorentz's 1895 theorem of the corresponding states. Einstein failed to make the leap from Lorentz's theorem to actually claiming or proving covariance.

When reasoning is given for crediting Einstein, it is usually backwards. It might credit Einstein for abolishing the aether, or for accepting the realism of the transformations, or for having the simplest postulates, or for explaining the physics, or for not being ad hoc, or for unifying space and time. But all of these arguments reverse the facts of what actually happened, and are reasons against crediting him.

The history of special relativity does not start and end with Einstein's postulates, as some books say. It started with a belief in the motion of the Earth that dated back to Aristarchus, Kepler, and Newton. Maxwell's electromagnetic theory was the first truly relativistic theory, and it led him to suggest a way to test the theory by trying to detect that motion. The theory also showed that light was an electromagnetic wave. Experiments by Michelson and others failed to detect that motion. Lorentz interpreted them as saying that the speed of light is constant for all observers, deduced his transformations, and showed that they were consistent with Maxwell's equations. Poincare perfected Lorentz's theory, and reformulated it in terms of spacetime symmetries. Minkowski popularized relativity as a spacetime theory.

Crediting Einstein for special relativity requires believing that he had some superior understanding to what he actually wrote, and that Lorentz's and Poincare's understandings were somehow inferior to what they actually wrote. But we know that Poincare understood the spacetime geometry of relativity in 1905, because he presented it in two different ways. He did not just stumble onto a correct formula by accident. He gave the symmetry group and the metric tensor. And Einstein certainly did not understand it, and we know that because he was still resisting it three years later when Minkowski was promoting it. Likewise Poincare certainly understood electromagnetic covariance, because he proved it two different ways and corrected Lorentz's error. And Einstein was still struggling with covariance ten years later. Lorentz may not have understood the ramifications of what he said, but we know that Poincare did, because Poincare built on them. And we know that Minkowski understood Poincare, because Minkowski cited Poincare and extended his results.

The essence of special relativity is the 4-dimensional geometry of spacetime and the electromagnetic covariance, and Einstein had nothing to do with the discovery or early popularization of these concepts. Einstein's 1905 and subsequent relativity papers were not seen at the time as any great breakthrough, but merely a minor elaboration of the work of Lorentz and others. It was not until Minkowski's 1908 lecture was widely disseminated that relativity became regarded as a new theory of space and time.

The reader is likely to be baffled as to how Einstein could be credited with writing the greatest scientific paper ever written, when it was not even the best paper on the subject in the month that he wrote it. It is indeed baffling. But the papers are all readily available, and you can verify everything yourself if you wish. My personal perspective is that of a mathematician. Mathematicians pursue knowledge according to standards of progress and rigor that may seem incomprehensible to others. They place great value on novel concepts and rigorous proofs. They like to credit those who do the actual work.

It might seem odd that Poincare did not do more to claim credit for himself during his lifetime, but it is not odd. He knew that his arguments were sound, but not everyone understood them. He wrote that "the postulate of relativity may be established with perfect rigor." Einstein's papers are shallow and derivative by comparison. Poincare was worried that some experiment would prove him wrong. No experiment did, and Poincare's ideas turned out to be fundamentally important for 20th century physics.

Even if you accept that Einstein deserves credit for inventing relativity, there is no other example in the history of science for refusing credit to his contemporaries who had brilliant contributions. And yet they are routinely denigrated on obscure terminological and philosophical grounds.

Historically the development of special relativity was that Maxwell clearly enunciated the relativity principle and its apparent conflict with his theory of electromagnetism, Michelson did the crucial experiment and created a paradox, Lorentz showed how transformations of space and time could resolve the paradox, Poincare turned Lorentz's ideas into a modern theory founded on the non-Euclidean geometry of spacetime, and Minkowski popularized Poincare's theory. There were also contributions by Voigt, FitzGerald, Larmor, Einstein, Planck, and others, but they had little effect on how special relativity became an essential part of physics.

Why relativity is fundamental

Relativity has a popular reputation for being an esoteric subject of little practical significance. As it is usually explained, relativity is only important when velocities are close to the speed of light. Then it seems like a hindrance more than anything, because it spoils those science fiction stories in which spaceships have warp drives for traveling faster than the speed of light.

Testing relativity is not science fiction. Particle accelerators can make protons go more than 99% of the speed of light, but never faster than light. Atomic and hydrogen bombs release energy just as relativity predicts. Getting high accuracy on GPS positioning requires relativistic corrections to the clocks. There are some amazing experiments that can be done to eight or nine digits of precision, and relativity is needed for the theoretical explanation.

All of these relativity applications are important, but they do not quite capture why relativity was so important to 20th century physics. There are several stronger reasons for why relativity is so fundamental.

Relativity is what allows physical theories to respect causality. Causality is a fundamental premise of physics, like conservation of energy. Principles of thermodynamics say that there can be no perpetual motion machine, and principles of causality say that no time machine can take you back in time. Relativity imposes the speed of light as a speed limit on the propagation of matter, fields, information, or anything else. It prohibits action-at-a-distance. It allows theories in which events and actions cause things to happen at nearby locations and times, and can only cause a distant effect after a chain of local events is allowed to propagate.

Causality is so crucial to the scientific method that it is usually not even mentioned. Science has been enormously successful by reducing observation about the natural world to causes. Astrology, on the other hand, is not very amenable to scientific analysis because it has no theory about how patterns in the night sky could cause human behavior on Earth. The most successful scientific theories work by reducing observed actions to a chain of events, related by causes.

Poincare argued in 1902 that the aether was invented to preserve causality. So was relativity. The main motivation for Lorentz's 1895 relativity theory was explaining the motion of the Earth, and a large motivation for Poincare's 1905 relativity theory was to create a causal structure for spacetime. Particles, fields, radiation, and everything else are limited to propagating at the speed of light, and not instantaneously. This makes it possible to have mathematical theories for physics in which the future depends smoothly on the past.

Explaining causality also motivated the theory of electromagnetic fields. Maxwell was dissatisfied by theories in which particles acted on other particles at a distance. It made much more sense to him if the particles caused some sort of disturbance in some sort of medium, and if that disturbance then influenced other particles.[249] That justified the electromagnetic fields. His theory had relativity and light-speed causality built into the equations, although it was not fully appreciated at the time. Relativity and causality are so central to our understanding of electromagnetism that it is now hard to imagine any other kind of physical explanation of the subject.

Feynman's greatest accomplishment was to formulate a quantum theory of electromagnetism. It was a particle theory, with electrons and photons being the main objects. His rival theories were field theories. Relativistic causality was essential to all of the theories. They all explain interactions as matter perturbing empty space that is not really empty, and that perturbed empty space affecting other matter. They are all constrained by the constant speed of light and by Poincare symmetries.

The principle of conservation of energy is so important that new definitions of energy had to be invented just to preserve the principle. Likewise, causality is so important that the aether, relativity, and field theory all had to be invented to preserve causality. One reason that relativity is so fundamental is that it is essential to our understanding of causality, and causality is essential to our scientific understanding of the world.

There are some interpretations of quantum mechanics that seem to violate some notions of causality, so perhaps causality should be considered just another scientific hypothesis that is subject to falsification. Quantum mechanics is often explained as a *non-local* theory, where measuring one particle has an effect on a distant particle. However such experiments can also be explained in terms of causal physics.[250]

Even if causality were refuted somehow, it would still be an incredibly useful concept. Our best physical theories have causality built into the dynamical equations, and our current understanding of causality requires relativity.

Relativity is essential to understanding electromagnetism. Gravity is the most important force for cosmology, but electromagnetism is the most important force here on Earth. Relativity teaches that magnetism is a relativistic effect. More than that, the equations for electromagnetism require relativity, and we have no way to understand electricity and light without relativity. A detailed understanding of chemical bonds and other chemical properties of atoms requires relativity. For example, relativity is part of the explanation of why gold is yellow, why mercury is liquid, and how lead car batteries work.[251] While the quantum mechanics of particles was originally conceived as a non-relativistic theory, the extension to fields like electromagnetic fields is entirely relativistic. The term *quantum field theory* is synonymous with relativistic quantum mechanics. It explains why the electron is the universal unit of electric charge, as it is the observable portion of a quantum field that obeys universal laws. It explains how particles can get created and annihilated in the aether. Relativity transforms quantum mechanics from a particle theory into an aether theory.

Our best electron theories have nearly always been relativistic. The electron had been conjectured for a long time, but the first really good evidence for it was discovered in 1896. At that time, the dominant theories for it were Maxwell's and Lorentz's, and they were fully relativistic. A non-relativistic quantum theory for it was proposed in 1926, but it was made relativistic in 1928 by Dirac. So our understanding of electrons has always been relativistic, except for two years.

Relativity helps to explain the stability of matter. Relativity is used to explain how large stars can collapse into black holes, but that is not really so remarkable. It is much harder to understand why the ordinary matter in front of you does not collapse. A rock or any rigid body is a conglomeration of particles and fields that somehow coexist in a stable way. You might expect the like charges to repel each other, and the unlike charges to attract, and it is hard to see how any theory would explain all those forces balancing out to give a stable rigid rock.

Because of relativity (and electromagnetism and quantum mechanics), all particles are divided into *bosons* and *fermions*. The bosons are described by vectors, and the fermions by spinors. Identical bosons have a symmetry that allows them to occupy the same space, like photons in a laser beam. Identical fermions have an *anti-symmetry*, and cannot. Matter is made out of fermions – quarks and electrons. An atom has a nucleus made up of quarks, and of *orbitals* filled with electrons. The atom has structure and occupies space because of a fermion anti-symmetry that keeps the particles apart. Similar principles give structure to molecules of atoms, and to crystals and other rigid bodies. There is no other understanding of solid objects. Relativity explains the distinction between fermions and bosons, and how stable matter can be made of fermions, while energy is made of bosons.

Relativity made the study of symmetry groups an essential part of theoretical physics. Symmetries had been used before relativity. For example, physicists might solve a problem by first looking for solutions having rotational or other symmetries. But relativity opened up a whole new kind of symmetry – a symmetry between space and time. Soon after relativity was discovered, many other important symmetries were found. Electromagnetism has a circular symmetry that is separate from space and time. Symmetries were shown to be closely related to conservation laws, so conservation of energy and momentum could be seen as byproducts of symmetry groups. Symmetry groups were crucial for the development of quantum mechanics in the 1920s. When particle accelerators discovered hundreds of new particles, symmetry groups were used to organize and classify them. The standard model of particle physics uses the symmetry group as its central guiding principle.

Relativity finally resolved the mystery of the motion of the Earth. For millennia scholars grappled with the basic scientific question of whether the Earth moves or not. By 1900 it was clear that the Earth moved, but not so clear what exactly that meant. Relativity showed that motion is relative and explained those puzzling experiments like Michelson-Morley. Astronomers can measure absolute velocity by comparing to the cosmic microwave background, but the laws of physics can be written so that they are valid in any reference frame. Relativity clarified the meaning of space, time, and motion.

Relativity is often presented as a theory that is unsettling and counter-intuitive. But the world would be more incomprehensible without relativity. We need relativity for causality, and we need causality for a scientific worldview.

It is often argued that the theory of quantum mechanics was the most important breakthrough for 20th century physics. It is a much bigger and more complex theory than relativity, and it required a much different methodology. It has many more useful quantitative predictions, such as those used to make common electronic devices. But quantum mechanics is also more mysterious. It leaves you wondering whether electrons are particles or waves, and whether the universe is deterministic or probabilistic. Relativity neatly answers all of the questions it raises, and puts them all in a beautiful geometric structure. Relativity has become our bedrock understanding of space and time, and by 1910 there was no going back to previous notions.

4. Discovering the motion of the Earth

Special relativity was discovered from attempts to observe the motion of the Earth. But it was not the beginning or the end of the story. Trying to understand the motion of the Earth has puzzled some of our greatest thinkers for over two millennia. Resolving the issue led to new symmetry principles that became fundamental to 20th century physics, and the subject became increasingly sophisticated. Here is an outline of some of the major ideas.

Ancient astronomers discover periodic motion

The earliest science to study periodic phenomena was astronomy. The ancients knew that the Sun rose and set on a daily basis, and the seasons changed on an annual basis. The Moon was on a monthly cycle. The stars would rise and set each night, and the planets were on much longer cycles. Eclipses also showed repeating patterns. The ancient Babylonians, Greeks, and others developed models for these astronomical objects and events.

Predicting the orbits of the planets is the trickiest. The first obvious motion of the planets is to rise and set with the stars. The second is that they moved progressively against the background of the stars, with a period that is different for each planet. Thirdly, sometimes a planet will reverse itself and go backwards in what is called *retrograde* motion. These motions repeated themselves after regular intervals. The ancients visualized the planets and stars as being attached to *orbs,* which were gigantic spherical shells surrounding the Earth.Pythagoras (ca 500 BC) was a Greek philosopher and mathematician who is remembered today for the Pythagorean theorem about the lengths of a right triangle. He had a cult following of Greeks called Pythagoreans. They believed in using mathematics to find truth and harmony in nature. They also believed that the Sun was the center of the universe and the planets followed circular orbits.

The Greek philosopher Plato (ca 400 BC) proposed explaining astronomy with uniform circular motion. It was easy to see how the Sun and Moon could be moving in circles, but Mars had retrograde motion. Circles don't change direction. His student Aristotle (ca 350 BC) considered the arguments for and against the Earth's motion, and decided that the Earth was stationary. He said that we don't feel the motion of the Earth, and we don't notice the birds getting left behind by a rotating Earth. We don't feel the centrifugal forces that usually send things outward from spinning objects. We also don't notice any *parallax*, or change in the pattern of stars in the night sky between summer and winter, as one might expect if the Earth was going around the Sun.

Plato and Aristotle had somewhat different philosophies. Plato was a big believer in abstract ideas, some of which were guided by idealism and aesthetics. He would not necessarily switch to a new theory because of some observations, because there could be other theories that explain those observations. Aristotle was much more of a realist and taught that observations could scientifically tell us how the world really is. So Aristotle was more likely to leap to conclusions from empirical data, while Plato believed that reason was the highest form of truth. There is no agreement even today as to which philosophy is better.

They understood that there is a distinction between reality and observation. Plato told a parable about people trapped in a cave where they could just see shadows. They would think that the shadows were real, until they escaped and discovered what was causing the shadows. The enlightened philosopher has a hard time convincing people that the shadows are just projections of some grander reality.

The Greeks were not stupid. They had legitimate arguments about motion, even if they were fallacious. Aristotle had correctly explained that the Earth was round based on eclipses and on seeing different stars in different latitudes. Later, Greeks would cleverly use geometry and astronomy to estimate the diameter of the Earth to within 10%. They noticed that the Sun was higher in the sky at noon if they traveled a couple of hundred miles south. By measuring the angles to the Sun, they deduced what would be later called the latitude. From estimates of the distances between lines of latitude, they could estimate the circumference of the Earth.

Aristarchus (310-230 BC) and other Greeks found clever methods for determining the distance to the Moon, and to the Sun. During a lunar eclipse, the Earth's shadow appears to be about four times the width of the Moon, so they figured that the Moon's diameter was one fourth of the figure that they already had for the Earth's diameter. By measuring some angles during a half Moon and a solar eclipse, Aristarchus was able use the sizes of the Earth and Moon to get an estimate of the distance to the Sun, and the size of the Sun. Experimental error caused his estimates to be inaccurate, but he correctly deduced that the Sun was a great many times larger than the Earth, so it made more sense to him that the Earth went around the Sun.

The idea behind parallax is simple. With two slightly different views of an object, you can estimate its distance. It is how you judge distances with your binocular vision. Your two eyes give images, and your brain does the rest. Mathematically, it is called triangulation because it uses the fact that a triangle is determined by the length of one side and two angles. Your brain knows the distance between your eyes and the angles at which each eye perceives an object.

Parallax fails if the object is too far away. You can easily judge distances within 100 feet, but when you look at the night sky, you are unable to see any depth at all. It is like watching a movie without 3-D glasses. Ancient astronomers knew that they could travel as much as a 1000 miles on Earth, and not notice any change in the sky patterns. There were, of course, all the daily and seasonal changes, but no changes that could be attributed solely to the distance between the viewing points.

The lack of any observable parallax caused all sorts of strange ideas in ancient astronomers. They just didn't know if all the objects in the sky were at the same distance from Earth, or at all different distances. They did not know whether they were distinct objects. Maybe the stars were all just pinholes in a gigantic spherical shell.

We now know that the pattern of the stars does not appear to change because the stars are very far away, and you need telescopes to detect the changes for even the nearby stars. Such changes were discovered in the 1800s. The Foucault pendulum gives a pretty convincing argument for the rotation of the Earth, but that was not discovered until 1851. The pendulum is a heavy weight on a long string, and is allowed to swing back and forth. The swinging appears to slowly rotate over the course of a day because the Earth is rotating underneath. By then, the motion of the Earth seemed obvious. But this information was not available to the ancient Greeks or to medieval astronomers.

Thus the Greeks had some pretty good astronomy, without being sure whether the Earth moves. And their knowledge was limited for some good reasons.

Ptolemy was a second century Greek/Roman/Egyptian who wrote great treatises on astronomy, geography, astrology, and other subjects. His astronomy book *Almagest* had a mathematical model of the Sun, Moon, stars, and planets, and was based on theories and observations that had been collected for centuries. His geography book had an atlas with maps of the known regions of Earth. He had access to a wealth of ancient wisdom. His works were preserved by the Arabs and used for well over a millennium.

Of course, his maps were wildly inaccurate by today's standards, and he did not even know about the Americas. His Earth circumference was about 30% less than the true value, which would later give some encouragement to Columbus. But he did know that the Earth was round and yet he drew flat maps anyway. It is geometrically impossible to accurately represent a round Earth on a flat map, and yet flat maps are extremely useful.

The curious thing about Ptolemy is that everyone faults him for his geocentrism in the Almagest, but no one faults him for his flat maps or even his astrology. Apparently people are smart enough to understand that flat maps have utility, but not smart enough to understand that a geocentric model might have utility. For example, the American astronomer Carl Sagan blamed Ptolemy for blocking progress:

> This model permitted reasonably accurate predictions of planetary motion, where a planet would be on a given day, certainly good enough predictions for the precision of measurement in Ptolemy's time, and much later. Supported by the Church during through the Dark Ages, Ptolemy's model effectively prevented the advance of astronomy for 1500 years.[252]

It was Islamic scholars who preserved the Almagest, and translated it into Arabic. Even the title is from the Arabic, not the Greek. All subsequent progress in astronomy was based on Ptolemy. The main thing holding up progress was the lack of technology to make more precise observations.

Here is the introduction to Ptolemy's astrology treatise, where he starts by explaining the difference between astronomy and astrology:

> Of the means of prediction through astronomy, O Syrus, two are the most important and valid. One, which is first both in order and in effectiveness, is that whereby we apprehend the aspects of the movements of sun, moon, and stars in relation to each other and to the earth, as they occur from time to time; the second is that in which by means of the natural character of these aspects themselves we investigate the changes which they bring about in that which they surround. The first of these, which has its own science, ... [253]

He goes on to explain that astronomy, as described in his own Almagest, is much more scientific than astrology. He argues that we should not dismiss astrological prognostication because it is sometimes mistaken, because we do not dismiss physicians and others for their errors. The point here is that he defines astronomy as how we view the Sun, Moon, stars, and planets *relative* to the Earth, and that is what he claims to be scientific. (He uses the word "stars" to include both stars and planets.) Whether the Earth moves or not, Ptolemy was completely correct in saying that his book describes the celestial movements relative to the Earth.

You might think that Ptolemy should be discredited for even showing interest in something as unscientific as astrology. But such interest was common among astronomers. Even many centuries later, the great scientist Galileo seemed to have a similar attitude towards astrology, and practiced astrology all of his life.

The Greeks had already understood, centuries earlier, that a similar model could be given for a Sun-centered system. Archimedes mentioned it, four centuries before Ptolemy:

> But Aristarchus of Samos brought out a book consisting of certain hypotheses, in which the premises lead to the conclusion that the universe is many times greater than that now so called. His hypotheses are that the fixed stars and the sun remain motion less, that the earth revolves about the sun in the circumference of a circle, the sun lying in the middle of the orbit, and that the sphere of the fixed stars, situated about the same center as the sun, is so great that the circle in which he supposes the earth to revolve bears such a proportion to the distance of the fixed stars as the center of the sphere bears to its surface.[254]

Archimedes was estimating how much sand would be needed to fill the universe, as an exercise in calculating large numbers. He correctly pointed out that the universe must be absurdly large in a Sun-centered system, because no parallax of the fixed stars had ever been observed. The parallax was observed in the 1800s, and in the early 20th century astronomers showed that there were other galaxies that make the universe very much larger than Archimedes imagined.

Ptolemy does give some terrestrial arguments in the Almagest for the idea that the Earth is stationary, and he acknowledges that others hold a different view. If the Earth were rotating, he says, then it would be going very fast and it seems as if the clouds would get left behind. A bird might not find its way back to its nest. In a later treatise, *Planetary Hypotheses*, Ptolemy proposes an interpretation of the Almagest. He proposes that the Earth is at the center of a nested set of orbs, with one for each heavenly body. He gives an order for the planets, and estimates distances to the Sun, Moon, planets, and stars. But it was the Almagest that Ptolemy claimed to be scientific, it was the Almagest that accurately described the movement of the Sun, Moon, planets, and stars relative to the Earth, and it was the Almagest that lasted for a millennium. The Almagest did not depend on any hypotheses about planetary distances or Earth motion. It gave a scientific description of the appearance of celestial objects relative to the Earth.

Epistemologically, the Almagest was similar to Einstein's 1905 relativity paper. Both were concerned with the kinematics, not the dynamics. That is, Ptolemy was concerned with measuring the motion of objects in the sky, and not the forces causing that motion. Einstein was concerned with measuring the contraction of moving objects, and not the forces causing that contraction. Both were operationalist, and only tried to give formulas for observables, without trying to explain what was really going on. And both were widely misunderstood by those who later tried to infer more meaning that what was actually in the text.

The most complex part of Ptolemy's Almagest was his model of the planets. The apparent orbit of each planet was determined by two circles. For example, the view of Mars from Earth is computed from two circles, which can be interpreted as how Mars and Earth circle the Sun. In Ptolemy's terminology, these circles were called the *deferent* and *epicycle*.

The Moon's orbit is a good example of an epicycle. The Moon has a (roughly) circular orbit around the Earth, and the Earth has a circular orbit around the Sun. The orbits are closer to being ellipses, but circles are accurate enough for the ancient data. You can also consider the Moon to be orbiting the Sun, and then the orbit is best described with an epicycle. From the view of the Sun, the Moon orbits in a roughly circular way like the Earth, but the Moon appears to have an additional back-and-forth motion. For half of each month, the Moon is going faster than the Earth, and passes the Earth. For the other half of the month, the Moon slows down and is passed by the Earth. That back-and-forth motion is the epicycle. Mathematically, it means that the Moon's orbit around the Sun is described as a main circle (the deferent, matching the Earth's orbit) plus an additional circle (the epicycle). There is no simpler way to describe the Moon's orbit.

Ptolemy's epicycles for the other planets follow the same principle. Each planet orbits the Sun along a circle. From the Earth's frame of reference, the Sun appears to orbit the Earth. The orbit of Venus, for example, appears on Earth to be the combination of two circles, one for the orbit of Venus around the Sun, and one for the apparent orbit of the Sun around the Earth. Ptolemy did not attempt to interpret these circles as separate orbits, and just called the larger circle the deferent, and the smaller circle the epicycle. This use of epicycles is necessary any time a system has two circular orbits. It is just a mathematical expression of the combination of those circles.

Even today, using two circles is the simplest and most direct way to explain the orbit of Venus, or any other planet, as seen from Earth. If you approximate the orbits of Venus and Earth as circles, then the relative view of Venus from Earth is obtained by subtracting those two circles.[255] A relative calculation is nearly always done with a subtraction. A vector from one point to another point is obtained by subtracting the points. That is how NASA calculates spaceship trajectories today, and that is how Ptolemy calculated orbits in the Almagest.

The Almagest did not require accepting one particular mathematical method, as it sometimes gave alternatives. Instead of the deferent-epicycle method, it showed that the same results could be obtained from a *moving eccentric* method. Ptolemy is sometimes criticized for requiring a belief in epicycles, but in fact the Almagest only used epicycles as one possible way of doing a planetary computation.

A further complication for Ptolemy was that the planets do not move with uniform velocity. There was an *equant* point for each planet, and the planet goes slower as it is closer to the equant point. This was supposedly one of Ptolemy's own innovations, as there is no record of equants being used previously.

The most complicated planetary orbit was Mercury. The Almagest had Mercury going around a moving eccentric, giving it an elliptical orbit. As was later discovered, Mercury does in fact have the most elliptical orbit of the planets, with Mars being the second most elliptical.

The Almagest treated each planet independently. It did not say anything about the distances to the planets, and it would give the same results regardless of the order of the planets. It did not even attempt to describe what those planets were really doing, or how far away they were, or how they might appear from any other perspective, or anything like that. As a mathematical convenience, it assumed that all the planets had a mean distance of 60 units from Earth. It just predicted the apparent motion of the planets in the night sky, as seen from Earth. For the purposes of his calculations, it did not even matter whether the Earth was moving or not. He was just modeling the appearance of the celestial objects from the Earth. It is possible to deduce the order of the planets, and even the relative distances, from the sizes of some of the circles in the Almagest. Saturn was the farthest known planet, and it had the largest deferent compared to its epicycle. But Ptolemy regarded such deductions as untestable hypotheses, because there was no known way of directly measuring the distances to the planets. If parallax could be observed, then the distances could be estimated with triangulation, but the planets were too far away for that, using the technology of the day.

Ptolemy was just explaining the 2-dimensional image of what you see in the sky, as if you were watching a giant spherical projector screen on top of the atmosphere. The Almagest is sometimes criticized for exaggerating the variation in the distance between the Earth and the Moon. But Ptolemy was not attempting to model such distances. He catalogued the brightness of a thousand stars, but he could not explain the variation in the brightness of the planets. That brightness depends on how close the planet is to the Earth, and how directly it reflects sunlight. He did not model such things. His shortcomings were not that his explanations were wrong, but that they left many things unexplained.

Visualizing Ptolemy's model seems strange today. You might wonder what happens when an epicycle intersects the orb of another planet. The model makes no attempt to prevent such disasters, or to give any intrinsic meaning to the orbs and epicycles. It does not even show any diagrams of the universe, and just shows diagrams of how particular orbits are viewed. That is because the Almagest was not supposed to be a realistic cosmological model of the universe. The science of astronomy was limited to methods for calculating the appearance of the sky. The cosmology of what was really going on out in space was another subject. The subjects remained separate until Kepler tried to combine them over a millennium later.

Nowadays it is common to say that Ptolemy was wrong to say that the Sun revolved around the Earth, but he accurately described relative movements in his Almagest. Ptolemy's critics are the ones who are wrong to take his model too literally. Even today, planetariums are built with a geocentric system, and nobody says that they are wrong. The planetarium shows you visually an image of the night sky as seen from Earth, and that is what the Almagest did mathematically. Likewise, when an almanac gives times for sunrise and sunset, it is not making a wrong claim about the Sun's motion. It is simply predicting how you can observe a relative motion.

Medieval astronomers rediscover Earth motion

Ptolemy's Almagest was used for over a millennium, as ancient, Islamic, and medieval European astronomers updated his tables and used his system to predict the night sky. The next major astronomical model was published by Nicolaus Copernicus in 1543. The book was titled *De Revolutionibus Orbium Coelestium* (On the Revolutions of the Celestial Orbs), and it is one of the most famous science books ever published.

Copernicus is best remembered today for proposing that the Earth goes around the Sun. However that was not a new idea, and there was no resolution of the ancient Greek arguments for and against the motion of the Earth. Medieval astronomers were much more interested in Copernicus's tables and methods, as they gave an alternative to the Almagest.

Copernicus also used the Earth's orbit to relate the sizes of the orbits of the other planets. He was able to deduce the approximate distances of the planets from the Sun. Aristarchus had been able to estimate the distance from the Earth to the Sun, and to the Moon, but the Almagest made no attempt to estimate distances to other planets, as it was just concerned with the appearance of the night sky. Ptolemy could do his calculations as if all the planets were the same distance from the Earth. An advantage of the Copernican view was that one could discuss relations between the planets.

The Almagest used two circles for each planet, one of which corresponded to the Earth's orbit. By using the size of the Earth's orbit, Copernicus was effectively able to estimate the radius of the other circle. That radius is the distance from the Sun to the planet. Copernicus was able to give the order of the planets. Starting with the closest to the Sun, the known planets were ordered as Mercury, Venus, Earth, Mars, Jupiter, and Saturn. Others had guessed this order, as this is the same order as the speeds of the planets in the sky.

Copernicus used epicycles and other mathematical tricks to fine-tune his model and improve accuracy. His use of epicycles was very different from Ptolemy's. Ptolemy's epicycles were large and corresponded directly to the (circular) orbit of the Earth or another planet. Copernicus's epicycles were small and numerous, and were ad hoc adjustments to fit the data.

The retrograde motion of Mars and the other outer planets has a natural explanation in the Copernican model. Here is an analogy. If you watch a car driving down the freeway, then it looks as though it is going forward from most viewpoints. The major exception is when you are in another car going faster and passing it on the freeway. Then the other car looks as though it is going backwards. Mathematically, the relative velocity is given by subtracting the velocities of the two cars, and the other car looks as though it is going backwards because the difference is negative and points backwards.

Likewise, the Earth moves faster than Mars, relative to the Sun. When Earth is aligned with Mars, then it is like a passing car, and Mars looks as though it is going backwards. It is as simple as that. The Almagest explains retrograde motion the same way, although the terminology is different. It doesn't talk about the motion of the Earth, but gets the same result by comparing the deferent and epicycle motions. The Almagest is like explaining the passing car analogy by just giving the subtracted velocities, and never addressing whether the observer's car is moving or not.

When De Revolutionibus was published, the main technical innovation was considered to be the elimination of the equant point.[256] Astronomers found his tables and methods useful whether they believed in heliocentrism or not. Almagest equants complicated calculations, so a system without equants was appealing. Some medieval astronomers hated the equant. Copernicus said that it violated the symmetry of uniform circular motion, and that systems using it were like a "monster rather than a man".

The equant was a clever system for conserving angular momentum. As explained earlier, the rotational symmetry of the solar system means that angular momentum must be conserved. That means that a planet must speed up when it gets closer to the Sun, and slow down when it gets farther away. In Ptolemy's system, the Earth was not exactly at the center of the planetary orbits, and neither was the equant point. The planets would slow down when they were closer to the equant point.

When Kepler later devised his model of the solar system, he used the Almagest more than Copernicus's book. Kepler's famous second law explained the varying speed of planets, and was extremely similar to Ptolemy's use of equant points. When Hooke and Newton later proposed their universal law of gravitation, Newton's strongest argument for it was that it implied Kepler's second law. Ptolemy's equant idea was one of the most important ideas in the history of astronomy.

Copernicus believed strongly in uniform circular motion, like Plato, and did not like the idea of planets speeding up or slowing down. Plate 4b of his book was captioned, "The axiom of astronomy: Celestial motions are circular and uniform or composed of circular and uniform parts." So he used extra epicycles in the hope that he could recover Ptolemy's accuracy without equants. This was more appealing to some people, but not more correct. The planets really do speed up and slow down.

An early draft of De Revolutionibus credited Aristarchus with proposing a heliocentric system, but that was omitted from the first few published editions. The book did credit a fifth century astronomer for saying that Mercury and Venus revolve around the Sun.[257] The idea of uniform circular motion was not new either, as that was the ancient Greek belief until the introduction of the equant. Even before Aristotle, Plato explained the heavens in terms of uniform circular motion. Aristotle even thought that circles were the most natural motions for objects in the aether. Copernicus also seems to have used some ideas from Persian astronomy, although historians are not sure how those ideas got to Europe.

Copernicus was not able to achieve any great increase in accuracy. One of the biggest inaccuracies in the Ptolemaic system was that the position of Mars in the sky was off by an angle of about 5 degrees every 32 years. Copernicus reduced this error to 4 degrees. By comparison, the diameter of the Moon is about half a degree, as viewed from Earth. But Copernicus had extra epicycles, and his model was not considered significantly more accurate overall.

Decades later, Galileo became an advocate for heliocentrism. He learned about telescopes and soon he was making his own and observing the night sky. He discovered moons of Jupiter, craters on the Moon, phases of Venus, and spots on the Sun. He also saw the rings of Saturn, but thought that he was seeing some sort of triple planet. He started arguing for the Copernican model as being proved correct by his discoveries. He considered himself to be a Pythagorean. His 1610 book was an exciting explanation of his discoveries, and was readable by those who did not understand the mathematical models of Copernicus and Kepler.

Galileo was a brilliant scientist but unfortunately some of his arguments were fallacious and he ran into trouble with his friends in the Catholic Church. He argued that since there are moons orbiting Jupiter, not everything orbits the Earth, and he correctly rejected Aristotle's argument that we would notice the motion of the Earth. That did not really imply Earth motion. He argued that the ocean tides were evidence of the Earth's motion and he refused to believe that they were related to the Moon. We now know that the tides are caused by the gravitational pull of the Moon and the Sun, and not the Earth's motion. For someone who is sometimes regarded as "the Father of Modern Science",[258] some of Galileo's arguments were not very scientific.

Galileo's discovery of Jupiter's moons is sometimes said to be what disproved geocentrism, because he showed that not everything was in a circular orbit around the Earth. But Tycho had already made a thorough study of the Great Comet of 1577 and subsequent comets, and correctly concluded that they were beyond the Moon and probably orbiting the Sun in non-circular orbits. These discoveries caused some new thinking about astronomy, but they did not necessarily support Copernicus, because he believed in circular orbits around the Sun. Galileo wrote a book in 1623 arguing that comets were an atmospheric illusion, like a rainbow.[259] He wrote that the physics of the universe is written in the language of mathematics, and that those who ignore the math are "wandering around in a dark labyrinth." He arrogantly attacked some Jesuit priests, and made enemies of them. Unfortunately, he was wrong about the comets.

The ancient Babylonians and others understood that the ocean tides are related to the positions of the Moon and the Sun. Some anthropologists think that African cavemen used the Moon in order to predict low tides for collecting shellfish 70,000 years ago. It could have been one of the earliest human discoveries, along with fire and spears. And yet Galileo got it wrong, and had to be corrected by Church officials.

Galileo's theory of the tides was based on the translational and rotational motions of the Earth. If the Sun is stationary and the Earth is revolving around the Sun, then the Earth is moving at a speed of 67,000 miles per hour. The Earth is also rotating, so a point on the surface could be moving as much as 1000 mph faster or slower, depending on whether it is day or night at that point. Galileo thought that this daily fluctuation in the speed of the ocean water caused the tides.

A caveman could see that Galileo was wrong. His theory predicts one high tide and one low tide each day, with the tidal times being the same every day. In fact there are two high tides and two low tides every day, and each day's tides are noticeably later than the previous day's. The tides are much more related to the Moon than to the Sun. Galileo persisted in his wrong beliefs even when Church officials showed him evidence of twice-daily tides from around the known world.

Galileo is credited with enunciating a relativity principle. He persuasively argued that the Earth could be moving without us feeling the motion, just as someone can be inside a large ship and not notice its motion. But it appears that he did not really believe what he was saying. If the Earth's motion were really undetectable, then it would not cause the tides either. His theory was inconsistent.

Galileo had to stick to his tidal theory because it was crucial to his argument that the Copernican model was correct. The relativity principle by itself only gives an argument that the heliocentric and geocentric models were equally acceptable, and did not show that one model was better than any other. In order to tell the Pope that he was wrong, Galileo needed his tidal theory.

Galileo had corresponded with Kepler, and they argued about the tides. Galileo wrote that he was seeking Truth with a capital T.[260] Kepler understood correctly that the ocean tides were caused by the gravity of the Moon, and wrote:

> If two stones were placed anywhere in space near to each other, and outside the reach of force of (other bodies), then they would come together ... at an intermediate point, each approaching the other in proportion to the other's mass. ... If the earth ceased to attract the waters of the sea, the seas would rise and flow into the moon ... If the attractive force of the moon reaches down to the earth, it follows that the attractive force of the earth, all the more, extends to the moon and even farther ...[261]

He conjectured that other such forces like magnetism[262] might explain the planetary orbits, not realizing that they could all be explained by gravity. He was suggesting that two millennia of kinematic astronomy be replaced with a dynamic theory. Hooke and Newton would figure out such a dynamic theory a few decades later. Galileo did not seem to understand or accept what Kepler had accomplished, and rejected it as mere astrological superstition.

In retrospect, Galileo's best argument was on the phases of Venus. Through his telescope, he could see that Venus was sometimes fully lit up like a full moon, and sometimes partially lit like a crescent moon. He had learned from Kepler that the Copernican model predicted such phases. A full Moon is very much brighter than a crescent Moon, so you might expect that the ancients would have noticed the varying brightness of Venus. But the differences in Venus's brightness are not so noticeable to the naked eye because Venus is fullest when it is farthest from the Earth, and Ptolemy made no attempt to explain the variation. It was hard to see how to get such a range of phases in the Ptolemy model.

But the phases of Venus do not really contradict the geocentric view. Ptolemy's model said nothing about the distance between the Sun and Venus, or how the Sun's light might shine on Venus. Kepler's prediction was really an argument that Venus revolved about the Sun, not that the Earth revolved about the Sun. Tycho's model from decades earlier had a motionless Earth and would still give the phases of Venus just like what Galileo observed. So there was still no proof that the Earth moved or that the Sun was stationary.

Galileo got the attention of the Roman Church, which was the astronomical authority at the time. It was in 1582 that the Pope correctly added ten days to the calendar after astronomical evidence showed that errors had accumulated over the previous 1500 years. Pope Urban VIII was a friend of Galileo, and encouraged him to teach his astronomical ideas, as long as he did not teach that heliocentrism had been proved and did not contradict Church teachings. He could give the arguments for and against heliocentrism. They even debated the issue in person. The Church also retracted its approval of De Revolutionibus until nine sentences could be corrected. As long as the book did not say that heliocentrism was certain, and stayed away from dubious theology, the Church approved it. The uncorrected version of the book continued to be widely available anyway.

Copernicus commented that since the stars appear fixed, they must be very much farther away than all the planets. He added:

> So vast, without any question, is the divine handwork of the Almighty Creator.[263]

This was one of the nine sentences to be corrected. It wasn't that the Church had any objection to fixed stars or vast space, but that he had injected some unapproved theological reasoning.

Galileo then wrote a book titled *Dialog on the Tides* that ridiculed geocentrism, and submitted it for Church endorsement. The Church authorities knew that his argument about the tides was wrong, and required him to remove the word from his title. He put the arguments for geocentrism mockingly in the voice of a character named Simplicio (simpleton).[264] Perhaps the Pope was expecting his own arguments to be better represented in the book.

Galileo was seriously misrepresenting the geocentrism argument. Many Church scholars subscribed to the Tychonic system, and Galileo had no valid argument that Copernicus's model was any better than Tycho's. Galileo's arguments were entertaining, but his book was a silly straw man attack on the Church because he only mentioned out-of-date astronomical models. He was put on trial and ordered not to teach that the Sun was stationary and at the center of the universe. Actually Galileo was only claiming that the Sun was at the center of the planetary orbits, and not at the center of the stars. Today, Galileo is considered a great hero and his book was recently ranked by *Discover* magazine as the fourth greatest science book of all time, just between Newton's great treatise and Copernicus's De Revolutionibus.[265]

It seems laughable today that the Pope would be considered an authority on astronomy. But he was enough of an authority in 1582 to redefine the calendar for Europe, and that calendar is the same one that we all use today. It is easy to understand how Church officials could have seen such authority as a good thing.

The Church faced many other serious issues during that period. The Pope also used his influence and authority to help organize a Christian naval fleet to defeat the invading Islamic Ottoman Empire at the Battle of Lepanto in 1571. Power struggles between the Catholics and Lutherans led to a devastating German war in the early 1600s. The wide availability of cheap printed books meant that authorities could no longer control the distribution of knowledge. Heliocentrism was a minor issue.

The whole Galileo story is contrary to our modern notions of academic freedom, but the Church was correct that Galileo had not really proven his case. The Church did not know that someday relativity theory would prove that geocentric and heliocentric views were both valid, but it did know that both could be used to predict astronomical observations. The story is not the grand conflict between science and religion that is told in popular myth. The Church was not persecuting scientists for their ideas. No other scientist was put on trial like Galileo. St. Augustine said as early as the year 408 that Bible stories should be treated as allegory if they conflict with science.[266] Both Galileo and his prosecutors were in explicit agreement with this principle. If Galileo had stuck to teaching heliocentrism as a scientific theory, he would not have had any problems. The literal interpretation of the Bible is a later phenomenon, and the mainstream Christian churches do not subscribe to it.

Galileo's ideas were not suppressed to any significant extent. His banned book was available on the black market, and within a couple of years, it was translated into Latin and republished in other countries. While he was forbidden to write any more books, he did in fact write a very important new book on the motion of falling bodies. It was published in The Netherlands, without the approval of the Church.

If the Church truly made some scientific error in its analysis of Copernicanism, then presumably that error could be found in those nine changed sentences. But no one addresses any such error. The most unsettling idea in the history of cosmology has been the big bang theory. While Einstein and many other scientists had difficulty accepting it, the Church did not. In fact the theory was discovered by a Belgian physicist and Catholic priest. Father Georges-Henri Lemaitre published a paper in 1927[267] that presented a relativistic model for the expansion of the universe, with an initial singularity at a finite time in the past. He also used observational data to estimate the expansion rate of the universe, and gave a value for what is now known as *Hubble's constant.* The American astronomer Edwin Hubble got additional observational evidence for the expansion.

There is a popular myth that the history of science is one of iconoclastic geniuses who had to rebel against stodgy authorities who tried to suppress their work out of dogmatic beliefs. An NPR radio broadcast said that all science was heresy for 500 years.[268] The prime example is always the anti-science character of the Roman Catholic Church, and the prime example of that is the trial of Galileo. The myth is completely false. The Church has a long history of pro-science activities. There is no example of the Church persecuting a scientist other than Galileo, and even the Galileo story is not the example that people think it is.

Anti-Christian propagandists have gone all the way back to Alexandria, Egypt, in the year 415, looking for a second example. That is the site of the great ancient library, and the female scholar Hypatia was murdered in a local political dispute that year. The 2009 movie *Agora* portrays her as an atheist who was on the brink of discovering heliocentrism and the Earth's elliptical orbit, when the Christian authorities felt threatened by scientific knowledge, and had her murdered and the library books destroyed. The director was inspired by some myths promoted by Carl Sagan in the PBS TV show *Cosmos*. As a statement about Christian anti-intellectualism, the movie is completely false.

The only other allegation of a persecuted scientist was Giordano Bruno, who was burned at the stake for heresy in 1600. He was notable for having speculated about an infinite number of worlds like the

Earth, each with a Garden of Eden, Adam and Eve, fruit of knowledge, and maybe a crucifixion. But he was a Catholic monk, not a scientist, and his heresy was to deny the divinity of Jesus Christ.

Copernicus's model was intriguing, but not really a major advance. The big progress came later.

The new astronomy

The period around the year 1600 was an exciting time for astronomy, just as 1900 was an exciting time for relativity. A bright new star appeared in the sky in 1604. It was an ordinary star (like our Sun) that exploded into a *supernova,* and it was almost as bright as Venus for a few weeks. It is called Kepler's supernova because the German astronomer Johannes Kepler wrote a book about it in 1606. It was a rare event, as we have not had a Milky Way supernova since then. There have been only a few in recorded history, with the main ones being the one that the Danish astronomer Tycho Brahe studied one in 1572, and others in 0185, 1006, 1054, and 1181. The Dutch invented the telescope in 1608, and by 1610 Galileo had published *Starry Messenger* with his observations with it. Kepler published his *New Astronomy* in 1609. These books demonstrated a 3-dimensional richness to celestial objects. Suddenly the heavens were alive.

Tycho invented instruments for making much more accurate astronomical observations, and then systematically collected data. He did not have a telescope, but he collected the best astronomical data that anyone has ever collected without a telescope. After a few years, he had, by a factor of ten, the most precise data ever recorded, and the shortcomings of the Ptolemy and Copernicus models were becoming apparent. He devised his own model in which the Sun and Moon went around the Earth, and the other planets revolved about the Sun. He looked for stellar parallax and brightness variation as evidence of the Earth's motion, and did not find any. He theorized that if the Earth went around the Sun, then the stars must be very far away, and that some stars could be as large as the entire Earth's orbit. His observational data on the planet Mercury became essential for the acceptance of general relativity three centuries later.

Tycho and Kepler read De Revolutionibus, but did not make much use of its mathematical models. Kepler's early heliocentric models used equants and resembled Ptolemy's more than Copernicus's. Kepler got some inspiration for creating new models from Copernicus, and from his reasoning, the details of his models did not resemble those of Copernicus at all. Copernicus used small epicycles and constant speed, and Kepler did not. Kepler perfected the work of Ptolemy, not Copernicus.

You might think that Tycho and Kepler would be an unlikely combination because they disagreed about the motion of the Earth. There are even allegations that Kepler murdered Tycho.[269] But their collaboration was actually one of the greatest in the history of science. Tycho proposed to Kepler that Mars had a non-circular orbit, based on years of data. When Kepler eventually published the *Rudolphine Tables* long after Tycho's death, it was really a joint work. The motion of the Earth was just a detail. The book was used to successfully predict a *transit* of Mercury in 1631 and a transit of Venus in 1639. A transit occurs when a planet comes between the Earth and the Sun, like a solar eclipse. The transits could be celebrated by the heliocentrists and the geocentrists, as the calculations and observations were the same. It seems strange to refer to the Earth's motion as a minor detail, but there are many examples in science of a theory being able to make predictions without any certitude of the underlying physical reality. Another example is quantum theory making predictions without any assurance that light is composed of waves or particles.

While Galileo was making discoveries with his telescope, Kepler was working out a much better model of the solar system. He was the first to say that the planetary orbits were ellipses, with the Sun at a focus point near the center. He figured out the distances of the planets from the Sun and how the planets change speed to conserve angular momentum. And he accepted the idea that the tides were caused primarily by the Moon, and not by the motion of the Earth. He titled one of his books *Epitome of Copernican Astronomy*, but he really had his own system that did not have much to do with Copernicus.

Kepler's theory is often falsely explained as replacing Ptolemy's epicycles by ellipses. For example, a 2005 New Scientist magazine letter said:

> [Ptolemy's Almagest] explained observed positions fairly well, but as the centuries went by into Renaissance times, more and more epicycles had to be added to explain the latest observations. The "dark force" of the epicycles was necessary to make an unquestioned theory work, but the reason for their existence

> was never explained. Even when Copernicus proposed putting the sun at the centre of the universe in the 16th century, this only reduced the number of epicycles needed from 80 to 34. It was not until Kepler's calculations in the early 17th century that it became clear the planets were actually travelling in ellipses.[270]

But it was Copernicus who added his epicycles to compensate for removing the equant, and Kepler put something similar to the equant back in his model. Kepler did not replace Ptolemy's epicycles at all. Kepler accepted the heliocentric interpretation of Ptolemaic epicycles as being consequences of the Earth's orbit. Ptolemy got the apparent motion of Mars by subtracting two circles, one of which was called an epicycle, and Kepler got the motion by subtracting two ellipses. Kepler got greater accuracy by using ellipses, but he never eliminated the function of the Ptolemaic epicycles.

Sometimes the argument is made that Ptolemy was unscientific because his deferent and epicycle were described as two separate motions with no individual physical significance, instead being combined into the smooth motion of an ellipse. But the deferent and epicycle circles really *are* two separate physical motions, and combining them does *not* give an ellipse. One circle corresponds to the motion of the Earth, and the other to the motion of the other planet. A modern description of relative motion might not be any different.

The Copernican epicycles were much more ad hoc than Ptolemy's, but they were still physically real. They were just coordinates that were mathematically convenient for periodic motion. Approximating periodic functions by circles has been essential to modern science for centuries. The German mathematician Carl Friedrich Gauss used such methods with many more circles to model asteroid orbits in 1805, and Lagrange used them to model lunar orbits before that. Even today, astrophysicists use mathematical constructions similar to epicycles when they model planetary orbits, and no one questions the reality or validity of such methods.

Kepler was the first to figure out a way to use parallax to measure distances to planets. Copernicus had deduced the radius of each orbit from Ptolemy's epicycles, but there was no way to check those values. Kepler's brilliant idea was to observe Mars, and then to observe it again exactly one Martian year later. Then Mars will be in the same place, but will look as though it is in a different place because Earth will have moved millions of miles. Thus he had two different views of Mars and could use triangulation to estimate the distance. This also worked for other planets, and he did not even have to assume that the orbits were circular, as Copernicus had.

An Italian priest published 77 arguments against the motion of the Earth in 1651.[271] Among an assortment of non-scientific arguments, he pointed out correctly that a rotating Earth would cause a cannon shot to appear to slightly deflect to the East or West. Nothing like that had ever been observed. The apparent deflection is now known as the *Coriolis effect* after an 1835 paper by the French mathematician Gustave Coriolis. Calculating the effect became essential for 20th century warfare.

Heliocentrism became universally accepted in the late 1600s. Romer used it to measure the speed of light in 1676. Stellar aberration gave more evidence of Earth motion in 1725. Even the Catholic Church stopped requiring those changes to De Revolutionibus in 1758. By the 1800s, there was overwhelming evidence. The Foucault pendulum showed that the Earth was rotating underneath a pendulum, making the pendulum appear to rotate over the course of a day. Cyclones turn oppositely in the northern and southern hemispheres for similar reasons. The spinning of the Earth was known to cause a bulge at the equator. Long-range artillery shells appear to curve because the Earth is rotating underneath. Astronomers finally detected that the appearance of nearby stars changed somewhat from summer to winter, showing that the Earth was revolving around the Sun.

In 1805, the English astronomer William Herschel deduced that the Sun itself was moving. He tracked the apparent motion of many stars, and found that some of their motions could be explained by the Sun moving at a particular velocity:

> Now since, according to the rules of philosophising, we ought not to admit more motions than will account for the observed changes in the situation of the start, it would be wrong to have recourse to the motions of Arcturus and Sirius, when that of the sun alone will account for both of them; ...[272]

Note that he had no way of knowing for sure whether the Sun was really moving. The motion of the Sun was just a convenient hypothesis for simplifying the apparent motion of the other stars. It was a matter of philosophy, not science. The whole Milky Way galaxy could be moving, for all he knew.

There was even evidence for heliocentrism from electromagnetic theory in the late 1800s. Maxwell's equations seemed to explain electromagnetism perfectly, and the popular interpretation was that there was some sort of fixed aether that was a universal medium for light. It seemed only logical to conclude that the Sun and the fixed stars were all motionless in the aether. It was no longer possible to argue, as some medieval astronomers did, that the choice between geocentrism and heliocentrism was just a matter of convenience.

All that was needed was some experiment that would show the Earth's motion relative to the aether. Scientists required something like the Foucault pendulum, but using electricity, magnetism, or light instead of a mechanical pendulum. In the 1880s, Michelson devised experiments for accurately measuring the speed of light, and for detecting the motion of the Earth through the aether. His experiments indicated that the Earth was not moving.

Physicists were not ready to go back to a geocentric theory. Some suggested that maybe the aether gets dragged along with the Earth, so that the Earth would not be moving relative to the aether. Michelson redid the experiment in greater precision with his colleague Morley in 1887, and it still showed that the Earth appeared to be not moving. Others did similar experiments, and got similar results.

Lorentz proposed an electromagnetic solution in 1892. If an object suffered a *length contraction* in the direction of motion, then the speed of light would be the same, and the apparatus would be unable to detect motion. FitzGerald made a similar suggestion, a couple of years earlier. Lorentz's paper was titled, "The Relative Motion of the Earth and the Aether", and the main point was to explain experiments that indicated that there was no such motion. This was the beginning of special relativity, and the idea that no optical experiment could detect the motion of the Earth.

The discovery of general relativity theory in the early 20th century made the Earth's motion even more meaningless. It became understood that the laws of physics are valid in any reference frame, if interpreted correctly. There was no longer a reason to think that the Sun was motionless. It was soon learned that the universe was composed of galaxies, the galaxies are rotating, and the whole universe is expanding. The Doppler effect can be used to measure whether stars have a velocity towards us, or away from us. Most of the galaxies are spreading out, but the Andromeda galaxy is actually on a collision course with our galaxy, the Milky Way. Geocentric and heliocentric reference frames are both valid, and either may be used depending on which is more convenient. Copernicus was no more correct than Ptolemy after all. Those who say otherwise are relying on physics that is a century out of date.

Comparing heliocentrism to relativity

Many philosophers and historians have compared the discovery of relativity to the work of Copernicus. According to some, these were the two biggest breakthroughs in the study of motion, if not the entire history of science. These comparisons started early. About special relativity, Planck said in 1910:

> In boldness, it probably surpasses anything so far achieved in speculative natural science, and indeed in philosophical cognition theory; non-Euclidean geometry is child's play in comparison. And yet the relativity principle, in contrast to non-Euclidean geometry, which so far has been seriously considered only for pure mathematics, has every right to claim real physical meaning. This principle has brought about a revolution in our physical picture of the world, which, in extent and depth, can only be compared to that produced by the introduction of the Copernican world system.[273]

The comment about geometry is odd, because the preferred understanding of relativity at that time was that it was a consequence of the non-Euclidean geometry of spacetime. If non-Euclidean geometry is child's play, then so is special relativity.

Planck was one of the most distinguished physicists in Europe, and Einstein's career benefited tremendously from this sort of talk. Weyl wrote a 1918 book on relativity, and said in the introduction, "This revolution was promoted essentially by the thought of one man, Albert Einstein." He incorrectly attributed several ideas to Einstein, and wrote:

> we are to discard our belief in the objective meaning of simultaneity; it was the great achievement of Einstein in the field of the theory of knowledge that he banished this dogma from our minds, and this is what leads us to rank his name with that of Copernicus.[274]

These analogies continue today. Ginzburg's autobiography says:

> The most radical innovative concept of the STR is that time is not absolute any more (this concept was put forward by Einstein). In its significance and intellectual challenge this concept can be compared to the rejection of the absolute immobility of the Earth, on which Copernicus built his heliocentric system.[275]

These premises and concepts were put forward by Poincare five years ahead of Einstein. This is even Poincare's terminology. Poincare's popular 1902 book explicitly said, "There is no absolute time." It is easier to read Einstein as believing in some sort of absolute time. His famous 1905 special relativity paper says that there is no absolute rest and says:

> It is essential to have time defined by means of stationary clocks in the stationary system ...

Einstein used the word "stationary" 62 times in the (translated) paper. His terminology was very similar to Lorentz's 1895 book, which repeatedly refers to stationary bodies and systems much more than it mentions the aether. Poincare never said that it was essential to define time by stationary clocks or aether clocks.

Copernicus also pretended to be banishing dogma from our minds by misrepresenting his sources. He blamed Christianity for belief in the flat Earth, and suggested that religious narrow-mindedness might reject his ideas. A draft of his book credited Aristarchus for heliocentrism, but that was omitted from the published edition. He appears to have made use of the *Tusi couple* to construct epicycles, without attribution.

The analogy to Copernicus was not made about Lorentz's 1895 theory or Einstein's 1905 theory. It was only made when it was clear that relativity was a theory about space and time, and not just a theory about electromagnetism. In his long 1905 paper, Poincare was the first to make this analogy between Copernicus and relativity. He proposed that gravity waves propagate at the speed of light, and wrote:

> Suppose, then, that this discussion is settled in favor of the new hypothesis, what should we conclude? If propagation of attraction occurs with the speed of light, it could not be a fortuitous accident. Rather, it must be because it is a function of the aether,

> and then we would have to try to penetrate the nature of this function, and to relate it to other fluid functions.

He is saying that it cannot be just a coincidence that light and gravity travel at the same speed. If so, the coincidence must be from a property of the underlying space (or spacetime, aether, or vacuum, depending on your choice of terms). Nobody has ever been able to measure the speed of gravity waves, but it is generally accepted that they do travel at the speed of light because of this argument from Poincare.

Poincare's 1905 papers both start with a statement about how all attempts to measure motion relative to the aether have failed. This gravity wave argument is the only other mention of the aether in those papers. It is sometimes said that Poincare had a theory that depended on the aether, but the aether plays no part in his theory. He only uses the term in peripheral remarks such as this.

He tries to explain the coincidence with an analogy to Copernicus:

> But the question may be viewed from a different perspective, better shown via an analogy. Let us imagine a pre-Copernican astronomer who reflects on Ptolemy's system; he will notice that for all the planets, one of two circles – epicycle or deferent – is traversed in the same time. This fact cannot be due to chance, and consequently between all the planets there is a mysterious link we can only guess at. Copernicus, however, destroys this apparent link by a simple change in the coordinate axes that were considered fixed. Each planet now describes a single circle, and orbital periods become independent (until Kepler reestablishes the link that was believed to have been destroyed).

He is saying that in Ptolemy's system, the apparent orbit of each planet is represented as the difference of two circles, one of which has a (sidereal) period equal to one Earth year. Ptolemy treats all the planets separately, so it is an unexplained coincidence that this same time period is used for each planet. Copernicus made a simple coordinate change to a Sun-based frame of reference instead of an Earth-based frame. As a result, all of those circles with the same period become identified with the Earth's orbit, and the coincidence is explained. Each planetary orbit is just one circle, and the view from Earth is obtained by subtracting the Earth's circle. Thus Copernicus explains the coincidence by taking a different view.

Poincare glossed over some details, because Copernicus actually used several circles for each planet. Each planet had one main circular orbit, which was corrected by smaller epicycles. For simplicity, Poincare is referring to just the main circles. By destroying the link, Poincare means that once the coincidental periods were removed and combined into one Earth period, the periods of the planets are unrelated in the Copernican model. Kepler reestablished a relation between the planets with his third law, saying that the square of a planetary period is equal to the cube of the distance to the center. Poincare goes on:

> It is possible that something analogous is taking place here. If we were to admit the postulate of relativity, we would find the same number in the law of gravitation and the laws of electromagnetism — the speed of light — and we would find it again in all other forces of any origin whatsoever. This state of affairs may be explained in one of two ways: either everything in the universe would be of electromagnetic origin, or this aspect — shared, as it were, by all physical phenomena — would be a mere epiphenomenon, something due to our methods of measurement. ...

Poincare wanted to explain the speed of light coincidentally being found connected to all forces, as he was proposing. One possibility is that all forces are electromagnetic. That was nearly Lorentz's view. His view was that the forces were either electromagnetic, or that they transform similarly. Lorentz was ahead of his time, as that was before it was realized that all chemical bonds and reactions are electromagnetic. It was usually correct for him to guess that some unknown force was electromagnetic. But Poincare was proposing a different view. He was saying that relativity was broader than electromagnetism, and was a property of our methods of measurement. That is, his new relativity was a theory about how we measure space and time.

Pursuing the analogy, Poincare was not exactly claiming that Lorentz was wrong:

> Perhaps if we were to abandon this definition Lorentz's theory would be as fully overthrown as was Ptolemy's system by Copernicus's intervention. Should that happen some day, it would not prove that Lorentz's efforts were in vain, because regardless of what one may think, Ptolemy was useful to Copernicus.

> I, too, have not hesitated to publish these few partial results, even if at this very moment the discovery of magneto-cathode rays seems to threaten the entire theory.

Poincare could be bold and modest at the same time. He declares himself to be the new Copernicus, and at the same time he acts as if he just has a minor improvement on previous work.

This passage makes it unmistakably clear that Poincare was presenting a new theory of the relativity of space and time, with a view distinctly different from Lorentz's. He was not clinging to old ideas about the aether; he was advancing something radical, original, and all-encompassing. He was proposing that relativity is a new theory about the measurement of space and time, and not just a property of electromagnetism.

When Minkowski proposed his version of relativity in 1907 and 1908, he also emphasized a geometric view of space and time that was distinctly different from the view of Lorentz, Einstein, and Planck. His Principle of Relativity was not a physical postulate, but a mathematical theorem about the covariance of physical laws on spacetime. Poincare and Minkowski were alone in this geometric view, but it quickly got other physicists excited after 1908, and it has been the dominant view in relativity textbooks ever since.

Curved spacetime

Understanding gravity requires a vast generalization of special relativity called *general relativity*. Just as the name implies, special relativity is a special case of general relativity. Special relativity is the linear theory, while general relativity is the nonlinear theory. Just as differential calculus teaches how to approximate curves by tangent lines, special relativity approximates a curved spacetime by linear tangents. Special relativity describes uniform linear motion with constant velocity, and general relativity describes acceleration.

Special relativity can explain electromagnetism because electromagnetism is linear. The magnetic field from two magnets is just the field from one plus the field from the other.

In modern terminology, the universe is a 4-dimensional manifold. That means that the vicinity of every point can be locally described by four real coordinates. Think of each point as an event, with three coordinates describing a spatial map, and one coordinate describing a clock.

A mathematician today might summarize special relativity by saying that spacetime is flat, meaning that it is composed of straight lines and the metric tensor is constant. General relativity teaches that spacetime is a (possibly non-flat) manifold with special relativity as a flat approximation. The metric tensor is a variable that varies from point to point, and event to event. The curvature is a measure of the changes in that metric tensor, and reflects the presence of matter.

The terminology is slightly confusing because special relativity has had important repercussions in many branches of physics, whereas general relativity has only been applied to certain esoteric cosmology questions. So it seems as if special relativity is the more generally applicable theory. Those branches need special relativity because it expresses the fundamental symmetries between space and time, and all of physics is concerned with space and time. General relativity also has those same symmetries, but the complications for extreme gravitational acceleration just are not needed for most of physics.

A simple thought experiment shows that acceleration leads to curvature.[276] Consider a simple disc. Euclidean geometry teaches that the circumference is 2π times the radius. Now spin the disc very fast. Special relativity now teaches that the circumference has a length contraction, while the radius is unchanged. The circumference is not 2π times the radius anymore. The disc is curved. It is similar to the way the Arctic is curved, because the Arctic Circle is less that 2π times the distance to the North Pole.

Another thought experiment shows that acceleration has some funny effects on time. Suppose that a spaceship has two clocks, one in front and one in back. The clocks would tick at the same rate if the spaceship had constant velocity, as the motion is not even detectable within the spaceship. But if it is accelerating, the surprising fact is that the clocks will run at different rates. The reason is that you could compare the clocks by moving the rear clock up to the front clock, but you would have to accelerate the clock to do it, and the relative motion causes a time dilation. The result is that velocity and acceleration must be considered when comparing clocks. Since acceleration is indistinguishable from a gravitational force, Einstein deduced in 1907 that gravity could also slow down clocks, just as uniform motion does. He later learned that gravity must also curve space.

Special relativity is based on the principle that the laws of physics are valid in any constant velocity. General relativity is based on a nonlinear generalization. It says that the laws of physics are valid in any frame of reference, whether accelerating or not. This principle is sometimes known as *general covariance*.

The nonlinear principle seems absurd. Newton's first law says that an object in motion will have constant velocity, unless a force is applied. In a constant velocity frame, such an object will still have constant velocity. The velocities will be different in different frames, but the object velocity will appear constant within each frame. In an accelerating frame, the object will appear to be accelerating. Newton only stated the relativity principle for constant velocity frames of reference.

If you were in a spaceship in uniform linear motion away from gravity, special relativity teaches there would be no experiment to tell you whether you are moving or not. The nonlinear principle does *not* mean that nonlinear motion would be undetectable in the same way that special relativity teaches that linear motion is undetectable. Nonlinear motion means acceleration, and you would feel the acceleration. But you will not be able to tell whether the acceleration you feel is from nonlinear motion or from a gravitational force. So in a theory that includes gravity, it is possible to say that nonlinear motion is undetectable because it can be detected as a gravitational effect.

In modern mathematical terminology, the relativity principle says that the laws of physics are well-defined on the spacetime manifold. They can be expressed independently of any particular coordinates or frame of reference. Or if expressed as equations in terms of some coordinates, it must be proved that the equations have the same meaning regardless of that choice of coordinates. If spacetime has been formulated as a manifold, then the term "covariance" is extraneous because only covariant equations even make sense on the manifold.

The mathematics of manifolds had been pioneered in the 19th century by the German mathematicians Gauss, Bernhard Riemann, and others, and by Poincare and Weyl and others in the early 20th century. The manifold is a subtle and difficult idea. Weyl explained it in a 1913 book titled, *The Concept of a Riemann Surface*. Once you accept this concept, it is natural to look for equations that are coordinate independent.

The general relativity field equations say that mass causes curvature of spacetime. If there is no mass, then spacetime is flat and the physics of special relativity applies. Once there is mass, there is curvature, and the whole notion of a straight line changes.

The surface of the Earth is roughly a sphere. It is curved. You cannot go in an ordinary straight line, because you have to follow the curves of the Earth. If you fly from London to Los Angeles, it will seem as if you are going straight, but you are really flying along a great circle having the same diameter as the Earth. In the spherical geometry of the Earth's surface, the straight lines *are* the great circles.

Light goes in straight line. In 1919, it was observed that starlight gets deflected by the Sun's gravity as it passes by the Sun. General relativity teaches that the light is still going straight, but the space (and spacetime) is curved. The light looks bent because the space is bent. Saying that a light beam is bent by gravity is a figure of speech like saying that the Sun sets in the west. The preferred explanation is that the Sun is not really setting, but the Earth is rotating to make it look like the Sun is setting. Likewise, the starlight is not really deflected, but spacetime is curved to make it look as if the starlight is deflected.

Just as objects tend to go in a straight line in the absence of forces, objects tend to fall in the presence of gravity. General relativity teaches that these principles are really the same thing. When a baseball player hits a fly ball, that ball is really going as straight as it can, but its trajectory looks curved because the Earth's gravity has curved spacetime.

Even more bizarrely, you could have a frame of reference centered at the fly ball. Then the ball does not go anywhere in space. It just goes forward in time. It is the baseball park that appears to be going on a parabolic trajectory. The equations for gravity are the same in any frame.

A frame centered at the ball may seem like a poor choice of frame, but there are no choices that make spacetime look flat. As long as there are masses present, then spacetime will be curved, and nearly all of the trajectories will appear curved no matter how you choose your frame.

At this point, you may think that I am presenting some goofy or nonstandard interpretation of gravity. But this is plain gravitational theory as it has been accepted for a century. Einstein himself explained it this way:

> Strictly speaking, one should not, e.g., say that the earth moves around the sun in an ellipse, since this statement presupposes a coordinate system in which the sun is at rest, ... In the investigation of the solar system nobody will employ a coordinate system at rest relative to the terrestrial body, since that would be impractical. But in principle such a coordinate system is according to the general theory of relativity fully equivalent to every other system.[277]

If you were an ant crawling around on that fly ball, you would not feel the force of gravity at all. You would be in free fall, like those weightless astronauts on a space-walk. It would feel like you were in an inertial frame. From that view, it is not so crazy to have a frame centered in the fly ball.

If you are sitting in a chair reading this book, it seems like you are not going anywhere. But you are going forward in time. You are moving in spacetime. In Newtonian lingo, the Earth is pulling you down, and the gravitational force is matched by a force from the chair pushing you up. The forces balance, and you don't go anywhere. In general relativity, being motionless means being in free fall. You are not in free fall. You only appear motionless because you are using an accelerating frame. You think that you are feeling gravity but you are really feeling acceleration.

In Aristotle's lingo, motion can be violent or natural. Natural motion is the motion of planets and fly balls that just follow the geometry of spacetime. Violent motion is motion caused by a non-gravitational force, such as a bat hitting a ball. Galileo echoed this distinction, but Newton did not.

All this sounds bizarre, but much of it is a consequence of the fact that gravity can be described in any coordinate system. Spacetime is a manifold with a metric tensor, and the consequences of gravity are all deducible from that metric, and the deduction can be done in any frame.

This is another example of how symmetry principles have guided 20th century physics. A theory of gravity had to locally have all the symmetries of a flat spacetime – translations, rotations, reflections, and

Lorentz transformations – and be invariant under nonlinear coordinate transformations as well.

Einstein deduced that Grossmann's gravitational field equations were consistent with explanations of the precession of Mercury's orbit. Others deduced black holes and the big bang theory.

A consequence of general covariance is that it is no longer correct to say that the Earth goes around the Sun. General relativity teaches that the field equations can be solved in an Earth-centered or Sun-centered or any other frame of reference. The same equations apply. The choice of coordinates is arbitrary.

In the late 19th century, there really was a rock-solid reason for believing that a Sun-centered view was correct, and an Earth-centered view was incorrect. Light was explained by Maxwell's equations, and those equations seemed to require an aether as a preferred frame of reference. It was believed that the aether was stationary with respect to the Sun and the fixed stars.

We now know that this 19th century view was mistaken. The Sun and stars are not fixed, and Maxwell's equations do not require a preferred frame. Anyone who expresses this view today is a century out of date. As the famous astrophysicist Fred Hoyle said:

> The relation of the two pictures [geocentrism and heliocentrism] is reduced to a mere coordinate transformation and it is the main tenet of the Einstein theory that any two ways of looking at the world which are related to each other by a coordinate transformation are entirely equivalent from a physical point of view … . Today we cannot say that the Copernican theory is "right" and the Ptolemaic theory "wrong" in any meaningful physical sense.

The modern view is that you can choose whatever frame is most convenient. For studying planetary orbits, a heliocentric (Sun-centered) system is usually preferred. This has the advantage that the Sun is (roughly) at the center-of-mass for the solar system. But if you want to study tidal forces on Earth, it is easier to use a geocentric (Earth-centered) system. Astronomy magazines usually give geocentric views for their most detailed charts, as that is what you see in the sky.

The 4-dimensional geometric view is that the Earth and Sun are both going in straight lines (also called *world lines*) because they are in free fall. Nothing stands still in spacetime because everything is moving forward in time. No event is ever repeated because it is impossible to relive the past. The Earth's orbit only looks curved because spacetime is curved. So someone could legitimately say that the Earth does not go around the Sun because both go in spacetime lines.

Before modern electronics and atomic clocks, relativistic time was not directly noticeable. Today you can buy a clock that will show an error if it is shipped on an airplane. Some research clocks are so accurate that they can detect the relativistic effect of weaker gravity from being raised a mere one foot, or of moving at a few feet per second.

General relativity also explains much of large-scale cosmology. The British Indian-American astrophysicist Subrahmanyan Chandrasekhar and others used it to predict black holes. Lemaitre, Friedmann, and others used it to predict the expansion of the universe.

It is hard enough to detect general relativistic effects at all. For decades, confirmation of general relativity was considered more qualitative than quantitative. Only recently has it been possible to do experiments that distinguish general relativity from rival theories.

The chief advantage to general relativity is that it abolished action-at-a-distance, and realized Poincare's vision of gravitational causality. It is no longer necessary to speak of forces from distant masses. Masses curve spacetime, and curvature guides motion.

The path to general relativity

Isaacson wrote that the 1915 general theory of relativity "was the product of a decade of solitary persistence during which Einstein wove together the laws of space, time, and motion".[278] No, it was not solitary. It was a product of mathematicians telling Einstein how to construct a nonlinear relativity theory.

The mathematics of general relativity is based on the spacetime metric tensor, combined with the curvature theory of Riemann, Ricci, Levi-Civita, and others. Einstein wrote a 1920 book on relativity where he describes the spacetime metric as the crucial breakthrough:

> These inadequate remarks can give the reader only a vague notion of the important idea contributed by Minkowski. Without it the general theory of relativity, of which the fundamental ideas are developed in the following pages, would perhaps have got no farther than its long clothes.

But what Einstein actually described was entirely contained in Poincare's 1905 paper. Poincare described the spacetime metric as being formally Euclidean if you use imaginary time, and that is exactly what Einstein praises in his 1916 book, without mentioning Poincare. [279]

Einstein was soon confronted with the spacetime metric from his old teacher Minkowski, who further developed Poincare's idea.[280] When Poincare's 1905 explanation in terms of a spacetime metric became popular in 1908, Einstein rejected it. By 1912, his old friend and mathematician Marcel Grossmann convinced him that it was essential to understanding gravity. By 1920, Einstein admitted that he would have been helpless without it.

Lorentz and Poincare took the first steps towards constructing a relativistic theory of gravity. Lorentz published an electromagnetic theory of gravity in 1900. Poincare declared that his principle of relativity applied to all laws of physics, and not just electromagnetism. He believed that this could be accomplished by constructing laws that are invariant under the Lorentz group. In 1905, he proposed how to make such a theory of gravity, and showed how gravity waves could propagate at the speed of light in a relativistic theory.

While writing a 1907 relativity review paper, Einstein found a way to apply relativistic time to gravity. He wanted to compare two clocks on Earth, with one on the ground and one falling. While it seems that the falling one is the accelerating one, it is the one on the ground is the one that feels the acceleration of gravity, and someone holding the falling one would feel weightless. By comparing gravitational acceleration to the acceleration of motion, he deduced a formula for how gravity could slow a clock. He later called this idea "the happiest idea of my life".[281] The effect was not measurable until many decades later, as clocks of the day were far too imprecise.

Einstein called his idea the *equivalence principle*, and described it as the idea that gravitational mass and inertial mass are the same. That was not really a new idea, as it had been conventional wisdom going back to Galileo and Newton. It is the reason that heavy objects fall at the same acceleration as light objects, ignoring air resistance, and it is the reason that planetary orbits can be calculated without knowing their masses. Poincare had assumed it in his relativistic theory of gravity, and had used astronomical evidence to argue that the two notions of mass agree to eight decimal places.[282] A Hungarian lab experiment showed similar results.[283]

Poincare looked for a way of experimentally testing a relativistic theory of gravity. He discovered in 1908 that anomalies in Mercury's orbit could be partially explained by relativity.

A fully-relativistic large-scale theory of gravity was going to have to explain how the spacetime metric in one area relates to the metric in other areas. For that, Einstein was going to have to learn differential geometry, a subject that was not studied by physicists of the day. Fortunately, his friend Grossmann was an expert on it, and tutored him.

The next step in constructing a relativistic theory of gravity was to figure what conditions on the metric would give a spacetime that resembled Newtonian gravity. Newton's law of gravity says that the force between two objects is a constant times the product of the masses, divided by the square of the distance.[284] Alternatively, the law can be expressed as the Sun creating a gravitational field in the surrounding empty space. The planetary orbits can then be explained as objects in a force field obeying the laws of mechanics.

Grossmann figured out in 1913 the condition for a relativistic gravitational field in empty space. Riemann and other mathematicians had shown that something called the *Riemann curvature tensor* was covariant and included all possible curvatures. Grossmann deduced that a certain combination of those curvatures, called the Ricci curvature tensor, must be zero in empty space. Grossmann published this in a joint paper with Einstein, but it was really two separate papers[285] as they did not agree on what the field equations should be. Grossmann was looking for covariant equations while Einstein was not. If a cloud of matter is present, Grossmann said that the Ricci curvature tensor was proportional to the stress-energy tensor of the matter. His equation had the virtue of being covariant, but was not quite correct in the presence of a cloud of matter.

It turned out that Grossmann's equation was all that was needed for the early applications of general relativity. Schwarzschild used it to find a solution in 1915 that was suitable for modeling the gravitational field surrounding a massive star like the Sun. This model sufficed to explain the precession of Mercury's orbit, the deflection of starlight, and the gravitational Doppler effect. These were the only experimental tests of general relativity for decades.

Cosmological models of the whole universe typically assume that the universe is a homogeneous and isotropic collection of dust. On a very large scale, the stars can be treated as dust particles. For these models, the full gravitational field equations are needed. These models became important when Lemaitre and others proposed the big bang theory in the 1920s. There is some controversy about the origin of those field equations.

Einstein did not accept what Grossmann had done in 1913. The Italian mathematician Tullio Levi-Civita urged Einstein in private correspondence to adopt general covariance for relativity, but Einstein did not believe that it was possible. In spite this advice and of advice from several other mathematicians, Einstein wrote several papers in 1913-1915 arguing that general covariance was impossible. In 1915 Einstein collaborated with Hilbert, who showed him how to derive the gravitational field equations in a coordinate-independent manner.

Hilbert was a very famous German mathematician and friend of Minkowski. Minkowski learned Lorentz's relativity theory in Hilbert's 1905 seminar on the subject. They discussed the contraction, local time, and Michelson-Morley, but not Einstein.[286] Part of Hilbert's fame is from announcing a list of 23 problems in 1900 that inspired a lot of 20th century mathematics. At some point he took up the task of making physics more rigorous, and announced, "Physics is too hard for physicists." He is primarily known among physicists for inventing *Hilbert space*, an infinite dimensional space that became the setting for quantum mechanics in the 1920s.

Most historical accounts of general relativity say that Hilbert wrote to Einstein with the field equations, and Einstein wrote back with an acknowledgement. Both submitted papers for publication with the field equations, with Hilbert's paper having the slightly earlier submission date. Einstein was very annoyed that he might have been scooped,[287] because he had been working on the problem for years and his joint paper with Grossmann had considered field equations very similar to what Hilbert had found. He was very upset that Hilbert might get credit for general relativity by just writing one paper on the subject that used years of previous work by others. Of course Einstein didn't mind getting credit for special relativity by writing just one paper on the subject.

The editor of Einstein's collected papers co-authored a 1997 paper[288] claiming that Hilbert could have gotten the field equations from Einstein, based on newly-discovered page proofs. The paper said that Hilbert revised his submission before publication, and alleged that the field equations were not in the original draft. They emphasized general covariance as the crucial idea and tried to give the entire credit to Einstein, but did not even mention Grossmann, Levi-Civita, and others. Those who subsequently examined those page proofs noticed that they had been mutilated, and that the portion that would have had the field equations is missing.

Regardless of whatever might have been on that missing half-page, the rest of Hilbert's draft gives a correct covariant formulation of general relativity. Hilbert correctly explains the necessity of using covariant expressions of the Riemann curvature tensor, something that Einstein had failed to understand when Grossmann introduced that tensor in their paper two years earlier. Hilbert correctly expressed the theory in terms of a covariant relativistic action. It is Hilbert's derivation of general relativity that is commonly taught in textbooks, not Einstein's. Hilbert's derivation was certainly original, as Einstein did not write anything similar. The only thing missing is a showing that Hilbert's formulation is mathematically equivalent to the one that Einstein subsequently submitted.[289]

The publication of the field equations by Hilbert and Einstein was treated as a great breakthrough, but only because Grossmann's equations had not been appreciated. The Hilbert-Einstein field equations were really just minor modifications of Grossmann's, and Grossmann's equations work fine for the solar system or any other situation where a gravitational field extends through empty space. The Schwarzschild metric, which gives the gravitational field of a star or a black hole, was developed from Grossmann's equations. In retrospect, Grossmann's equations also work fine in the presence of electromagnetic fields, and correctly predict how electromagnetic energy causes curvature and affects gravity. Even in the presence of matter, the only deficiency was that one of the ten equations needed to be corrected by the so-called *trace term*.[290] Grossmann would have very likely figured out that correction if he had known about the *Bianchi identities*, which are curvature properties known to the Italian differential geometers. Nine of the ten Grossmann equations were correct, even in the presence of a matter cloud. Hilbert's action approach gets the trace term correct. Einstein gave no explanation as to how he got the trace term.

It is unusual to deny someone credit because of a minor error. The biochemists James Watson and Francis Crick are popularly credited with discovering the double-helix molecular structure of DNA. Some people also give some credit to the American chemist Linus Pauling for developing the methodology that previously found helical structures in proteins, and to Rosalind Franklin since Watson has admitted that surreptitious use of her data was crucial to their work, and because she was the first to propose a double helix. But hardly anyone even mentions the fact that the published Watson-Crick model had an error that Pauling corrected.

Einstein's main general relativity paper was published in 1916, and it was written more like a textbook than a research paper. It credited Minkowski, Levi-Civita, and Grossmann at the beginning, but it did not cite any specific papers. It included a long explanation of tensor analysis, a subject that had been developed by mathematicians but was not well-known to physicists. It was not clear what was original.

While Grossmann's equations are sufficient for the solar system, large-scale cosmological models of the universe must take energy densities into account. Very little was known about how to test any such model. In 1998, it was discovered that the expansion of the universe was accelerating. This is best explained by hypothesizing a *dark energy* that permeates all of spacetime, like the aether. This requires adding a special vacuum energy term to the equations published by Hilbert and Einstein. As a result, the universe appears to be approaching *De Sitter space*, a model discovered in 1917 by de Sitter and Levi-Civita. It seems silly to argue about who first discovered the most general gravitational field equations when the necessary cosmological data was not known until decades later. The different possible field equations are all minor variations of the trace term in Grossmann's equations.

General relativity can be formulated in terms of an action principle, field equations, or a metric model. As it is usually explained, the action principle is used to derive the field equations, and the field equations are used to find a metric model. For applications like the solar system, the action principle was discovered by Hilbert, the field equations were discovered by Grossmann, and the metric model was discovered by Schwarzschild. That metric model is the relativistic equivalent of the Hooke-Newton inverse square law of gravity.

If general covariance was the crucial idea for general relativity, then we can be sure that the idea was not Einstein's. From 1913 to 1915, Grossmann, Levi-Civita, Hilbert, and possibly others tried their best to persuade Einstein of the idea, but he wrote papers saying that the idea was wrong. Sometimes when there is a question of credit, scientists are given the benefit of the doubt and credited with independent invention. But when one had published a great idea and another had published denunciations of the idea, then it is bizarre to credit someone for the idea that he denounced.

Thus the major breakthroughs to creating general relativity were Poincare's 1905 formulation of relativity in terms of a spacetime metric and covariant laws, and Grossmann's 1913 equations giving the gravitational field in terms of the curvature of the spacetime metric. Einstein persuaded various mathematicians to find relativistic gravitational field equations for him, and then he popularized the theory. For many years the field equations were known as the Einstein-Hilbert equations. Now they are just known as the Einstein equations, and Hilbert's action is known as the Einstein-Hilbert action. The applications of general relativity were done by many people.

The irrelevance of nonlinear relativity

While general relativity was hailed as a great intellectual accomplishment in 1919, it remained a fringe area of physics for decades. The linear part of the theory, special relativity, was well-confirmed and absorbed into other parts of physics. The nonlinear and gravitational effects were very difficult to test, and general relativity had little relation to mainstream physics.

The early tests of relativistic gravity were really just qualitative confirmations of the theory. The eclipse observations showed a starlight deflection that appeared to be greater than the Newtonian predictions, but that is about all. The test could not be done precisely until quasar observations in the 1960s.[291] The one test with quantitative agreement was the perihelion advance of Mercury. But there was just one number to compare, and it took centuries to collect the necessary data. There was no way to repeat the experiment. Some people looked for other explanations for the data, such as the oblate shape of the Sun or Mercury being tugged by an unknown planet *Vulcan*. Neptune had been discovered by studying similar anomalies in Uranus's orbit. There was no way to be sure that general relativity was the correct explanation.

Astronomy has long been understood in terms of cycles. Our day, month, and year are based on astronomical cycles. The phases of the Moon repeat every four weeks. The planet Neptune completes an orbit every 165 years. The orientation of the Earth's axis changes so that the signs of the Zodiac cycle every 26,000 years. Our solar system revolves around the center of the Milky Way galaxy every 230 million years. These can all be understood without relativity. General relativity teaches that Mercury's perihelion cycles every three million years (excluding effects from other planets).

When NASA sends probes to other planets, it ignores general relativity. Ordinary non-relativistic gravitational mechanics work just fine, as both methods give the same predictions to within the precision of the equipment. There are some anomalies where spacecraft acceleration has been slightly different from what gravity predicts, but these anomalies are present whether Newtonian or relativistic formulas are used.[292]

The use of curved spacetime is not as radical as was once thought. Cartan showed in 1922 that classical gravity could also be described in terms of curved spacetime.

The one practical application of general relativity is the Global Positioning System (GPS). This is a system of space satellites that allows receivers in cars and cell phones to show location on a map. The satellites need to have extremely accurate clocks because distance is measured by how far a radio signal can go in a measured time interval. Light travels at a speed of about one foot per nanosecond. A nanosecond is about the time it takes for a computer to add two numbers together, and an electronic clock can measure time in nanoseconds. So the clocks on the space satellites must be accurate to one nanosecond in order to get one-foot accuracy on the ground. They are actually accurate to about 40 nanoseconds, so we get about 40-foot accuracy on the ground.

But the space clocks run faster than Earth clocks, accumulating what would be errors of about 40,000 nanoseconds per day. More precisely, the net effect of 38,000 nanoseconds per day is obtained by combining the general and special relativity effects, from the Earth's gravity slowing down the ground-based clocks by 45,000 nanoseconds per day, and the faster motion slowing down the space clocks by 7,000 nanoseconds per day. That would give an error of about 38,000 feet (i.e., several miles) after one day of space clocks getting de-synchronized.

Everyone who tells this GPS story gives the impression that we could never have had GPS without Einstein and general relativity. But the story does not show that at all. General relativity gives a theory for gravitational fields curving space and time, but it is not really needed. Einstein published the formula for approximating how gravity slows down clocks in 1907, as a simple application of special relativity. That was before he learned about spacetime, and long before general relativity. GPS only uses the easiest part of relativity.

Furthermore, if we knew nothing about relativity, the GPS engineers still would have re-calibrated the clocks after launching them into space. They would have been mystified as to why the space clocks needed a 38 millisecond per day adjustment, but they would have done it anyway. As it is, they have to frequently update the clocks because of errors such as those caused by irregularities in the Earth's gravitational field. So they could have used kluges to avoid large errors, even without relativity. There would be a bunch of silly papers with speculation about what cosmic forces might be affecting atomic clocks, but we would still have a GPS system.

As before, it is possible to take the view that none of the clocks are really slowing down. The geometric view is that there is no such thing as gravity, and that satellites are traveling in straight line in a curved spacetime. The apparent slowdown is just an illusion caused by our poor ability to visualize the 4-dimensional geometry. We compare clocks in a way that seems reasonable in what we perceive to be a 3-dimensional world, but we are just seeing a slice of the true picture of events. As Poincare radically proposed in 1905, his preferred view of relativity was that it is "something due to our methods of measurement." To the GPS engineers, motion and gravity cause the clocks to slow down, but to a mathematician, these are just geometrical artifacts caused by the choice of the reference frame intertwining space and time.

Hundreds of millions of dollars have been spent in the last 20 years in attempts to detect gravity waves. The experiments use ultra-precise versions of the Michelson-Morley interferometer, except that instead of trying to detect motion of the Earth, they are trying to detect motion in a distant galaxy. So far, they have all failed.[293]

General relativity differs from classical mechanics by first order linear effects and second order nonlinear effects. The linear effects are well understood on every scale, from electrons to galaxies. The subject is made difficult by the nonlinear effects that are extremely small and incredibly hard to detect. They can only be detected on a large scale, such as tracking Mercury's orbit for centuries. Those nonlinear effects are fascinating to mathematicians and theoretical physicists, but are just irrelevant to almost everything we know about the world. Theoretical physicists like to complain that quantum mechanics does not apply to general relativity, but quantum mechanics deals with the sub-microscopic world of electrons and atoms. If something is big enough to see with the naked eye, then the quantum mechanical effects are negligible. The theories of quantum mechanics and (nonlinear) general relativity operate on domains that differ by many orders of magnitude. There is no way to even relate them, except maybe at some hypothetical first micro-instant of the big bang when all the known laws of physics break down anyway.

The search for consequences of the nonlinear effects leads astrophysicists to speculate about the center of a black hole. Stars with more than three times the Sun's mass can collapse after their fuel is spent. Once they get smaller than about ten miles in diameter, they become black holes and no light or other information can escape a boundary sphere called the *event horizon*. The theory says that the matter continues to collapse to a singularity, and the interior diameter becomes infinite, but there is no way to tell whether that is true or not. The theory also says that nothing inside the event horizon can influence anything outside, so the inside is not observable.

Making a relativistic quantum theory is technically difficult because it must have a symmetry making all velocities alike. Everything has to be the same if the whole system is going at close to the speed of light, where all sorts of funny things happen. Nonlinearities make the project more difficult. Gravity is slightly nonlinear because mass increases with velocity. Electromagnetism is simpler because the charge of an electron does not change no matter how it is moving. But if a gravitational object moves close to the speed of light, then the increased mass will generate stronger gravitational forces, complicating the interaction. Fortunately, the nonlinearities are negligible in all known situations. Planets do not go close to the speed of light, and electrons do not have much gravity.

It is widely believed that general relativity ought to be replaced by a *quantum gravity* theory. The thinking is that general relativity and quantum field theory are very different theories, and they have an inconsistency that must be resolved somehow. General relativity is large-scale, deterministic, and nonlinear, while quantum mechanics is small-scale, probabilistic, and linear. The speculation is that a perfect quantum gravity theory might explain the singularity of a black hole, or the initial instant of the big bang. But there is no known inconsistency that anyone would ever be able to observe in the foreseeable future. The interior of a black hole is not observable. There is no experiment that can be done to distinguish different quantum gravity theories.

All known physical examples of gravity can be quantized with something called *effective field theory*. Scientific American reports:

> "Everyone says quantum mechanics and gravity don't get along – they're incompatible," says John F. Donoghue of the University of Massachusetts at Amherst. "And you still hear that, but it's wrong."[294]

Nearly all aspects of relativity have been reconciled with quantum mechanics, including everything discussed in this book. The length contraction, time dilation, mass increase, magnetism, starlight deflection, solar system dynamics, and GPS are all fully relativistic and fully quantum mechanical, without any inconsistency.[295]

Quantum gravity research focuses on extreme nonlinear relativity, such as on the inside of a black hole. But quantum mechanics is founded on the study of observables, and the inside of a black hole is not observable. That is the real problem with quantum gravity. There is no expectation that any such research could have any bearing on cosmology or any imaginable experiment.

The standard model of particle physics

In the 1970s, the best theories of 20th century physics converged on a theory for all the known particles and forces. It is called the *standard model*. More Nobel Prizes have been given for it than anything else. With minor modifications, it is consistent with all known experiments. It is perhaps the most magnificent theory ever devised.

It is a theory of particles and fields. Like previous quantum field theories, you can think of it as a theory of particles that act like fields, or a theory of fields that act like particles. Or you can think of it as a theory of mysterious objects that look like point particles when you observe them, and like fields when you are not looking. It is not a theory of strings, like the much better known String Theory.

The only elementary particles are *quarks*, *leptons*, and *bosons*. The quarks are mainly the *up* and *down* quarks. A proton is composed of two up quarks and a down quark, and a neutron is one up quark and two down quarks. The leptons are mainly the electron and the *neutrino*, a tiny harmless particle that is emitted in radioactive decay. There are also other *flavors* of these particles, and also antiparticles. The bosons are the photon, particles that transmit other forces, and the *Higgs*. The Higgs is a conjectural particle that gives mass to the others, and doesn't do much else. It is named after the English physicist Peter Higgs, although others had some closely related ideas.

Bosons have the property that you can have lots of identical bosons in the same place. There is a symmetry that allows rearrangements of the identical particles, without any physical change. This is similar to what you might expect if you think of the bosons as tiny marbles or balls. The quarks and leptons are different. They are described mathematically with spinors, which means that 360 rotations introduce a funny minus sign. The mathematics of these particles is that a rearrangement of identical particles can introduce a funny minus sign, just like that 360-degree rotation. These particles requiring spinors are collectively called *fermions*. The Pauli exclusion principle says that identical fermions *cannot* be put in the same place.

Bizarre as this concept sounds, much of chemistry is based on the mathematics of spinors. A water molecule is formed when an oxygen atom forms bonds with two hydrogen atoms. A hydrogen atom has one electron. A pair of electrons can form a chemical bond because they can line up with opposite spin and fill an orbital. The oxygen atom has two unpaired electrons, and forms one such bond with each of the two hydrogen atoms. The chemical properties of water and other molecules can be understood from such considerations.

The upshot of these rearrangement symmetries is that fermions are good for building matter, and bosons are good for transmitting forces and energy. The book you are holding has substance because it is made of quarks and electrons, and those particles have spin-related symmetries that keep them apart. On the other hand, there is no substance to a laser beam, as unlimited numbers of identical photons can be concentrated in one beam.

The theory of the standard model builds on the success of quantum electrodynamics developed by Dirac, Feynman, and others. In modern terminology, it is a quantized gauge theory. Gauge theory was invented by Weyl who helped found quantum mechanics and who pioneered applying symmetry groups to physics.

Imagine that instead of 4-dimensional spacetime, we live in a 5-dimensional space that is just like spacetime except that there is an extra circle at every event. It is just an angle, from 0 to 360 degrees. Now imagine that there is also an extra circular symmetry in each such circle, so that each circle looks like just one event. Then the 5-dimensional space would look to us like the usual 4-dimensional spacetime, and we would not notice the extra dimension.

It seems as though nothing would be gained by adding a new dimension and a new symmetry that makes it invisible. It is not like an extra spatial dimension. It is more like a polarization or a phase that gains meaning when comparisons are made. There can be a curvature effect. Each particular circle has no physical significance, but a set of circles can have curvature, and the curvature can act as an electromagnetic field.

Curvature is a fundamental geometric notion. A surface can have curvature zero, like the flat plane, curvature positive, like a sphere, or curvature negative, like a saddle. If you were an ant living on the surface, you could detect the curvature by crawling in a small circle and comparing the diameter to the circumference. A flat plane obeys the laws of Euclidean geometry, and the ratio of the circumference to the diameter is π *(pi)*. The ratio is less than π when the curvature is positive and greater than π when the curvature is negative.

If every point has an invisible circle attached, curvature can be measured by going in a loop and comparing the starting circle point to the ending circle point. Whatever rotation takes one point to the other is a measure of the curvature in the area of that loop.

It turns out that electromagnetism can be understood in terms of the connections between these invisible circles. Maxwell's equations can be written in this way, with the curvatures being the electric and magnetic field strengths. When written this way, Maxwell's equations become truly relativistic. Not only are they invariant under Lorentz transformations, as proved by Lorentz and Poincare, but they are invariant under any nonlinear transformation.

Thus gravity and electromagnetism were unified in a pleasing way. The setting for both is a 4-dimensional spacetime manifold, a metric tensor, and connections between invisible circles at every event. The gravitational field is the curvature of the metric. The electromagnetic field is the curvature of the circle connection.

Quantizing the electromagnetic field turned out to be surprisingly difficult, but a very satisfactory theory was worked out by Dirac, Feynman, and others in the 1930s and 1940s. Poincare group invariance was a guiding principle.

A quantum theory of fields is always difficult for the following reason. The fields satisfy wave equations but are observed as particles. You can think of them as particles that look like waves, or waves that look like particles. Either way, the only way to equate a particle to a field is to say that the field has infinite strength concentrated at a point. If an electron is really just a point particle, then it has infinite density at that point. Worse, like charges repel, so it would take infinite energy to concentrate an electric charge in a point, so an electron should have infinite mass.

The only way to avoid these infinities is to have a system called *renormalization* to cancel infinities. Electrons can only exist in a sea of *virtual particles* that are necessary to cancel the infinities. You can call this sea an aether, or a vacuum state, or something else, but it has to be there in order for the theory to be renormalizable and to give non-infinite results.

Relativity makes renormalization particularly complicated. The vacuum is filled with pairs of virtual particles that briefly come into existence, and annihilate each other. In order for the theory to be symmetric under the Poincare group, the particles could have any velocity. Computing something as simple as an electron in a vacuum requires taking into account infinitely many virtual particles, and each such particle could have arbitrarily large velocity. The calculations involve infinities, and renormalization is a procedure for canceling the infinities.

The strong and weak forces appear to be nothing like the electromagnetic force. Physicists proposed many models for the strong and weak forces, but they all had the same problem. They were not renormalizable. The big breakthrough was when Dutch physicist Gerardus 'tHooft proved that gauge theories were renormalizable. This meant that the circle group of electromagnetism could be replaced by other groups to give consistent theories.

The weak force is explained as a gauge theory with the group SU(2). That is the group that looks just like the rotations in 3-dimensional Euclidean space, except that you have to rotate by 720 degrees to get back to where you started. Alternatively it can written as rotations in a 2-complex-dimensional space. Thus the universe has an SU(2) symmetry that is completely different from ordinary rotations.

We do not notice the SU(2) symmetry because it is broken. If it were a perfect symmetry, then the weak force would be transmitted by speed-of-light particles like the photon and would have long-range effects. It does not. The gauge theory equations for the weak force have the SU(2) symmetry, but the low-energy solutions do not.

Here is an example of a broken symmetry. The Sun's gravitational field has a rotational symmetry, so you might expect the Earth's orbit to be a circle. If the orbit were a perfect circle, then the orbit would exhibit that rotational symmetry. But the Earth's orbit is slightly elliptical. Since it is not exactly circular, the rotational symmetry is broken.

The mechanism for breaking the SU(2) symmetry is called the *Higgs field.* No one has ever seen a Higgs field. We have no reason to think that it exists, except that it is the simplest known mechanism for breaking the symmetry. There is solid experimental evidence that the symmetry exists and that it is broken, but we don't know for sure how it is broken.

If the Higgs hypothesis is correct, then there is a Higgs field everywhere in the universe. It is like another aether. All particles would be massless and travel at the speed of light like the photon, except that they appear massive if they interact with the Higgs field. There is some reason to believe that the Higgs boson will be discovered in a particle accelerator in the next several years. If so, it will be the most spectacular confirmation of theory in 30 years.

Gauge theory worked so well for the weak force that SU(3) was proposed for the strong force. That is the group of rotations in a 3-

complex-dimensional space. It is also a short-range force, but for different reasons.

Thus the standard model explains all the known forces as a gauge theory with the groups U(1), SU(2), and SU(3). All four fundamental fields are seen as curvatures. The standard model was worked out in the 1970s. At that time it was thought that neutrinos were massless and went at the speed of light. Since then it has been discovered that neutrinos have a small nonzero mass, and that the three flavors of neutrinos were more closely related than had been thought. Hence they go very fast, but do not quite go at the speed of light. The standard model was modified accordingly. There is also a better understanding of how quark masses relate to proton and neutron masses. Otherwise the standard model has been consistent with all numerical experiments since the 1970s.

The standard model is peculiar in that it is *not* invariant under a (spatial) reflection symmetry. The neutrinos are left-handed. There are actually some experiments involving weak interactions and neutrinos where the mirror image experiment would come out differently. The weak force just acts on left-handed fermions (and the corresponding right-handed antiparticles). This is startlingly non-intuitive as Newtonian physics and ordinary experiences have reflection symmetries. It is hard to even imagine a theory that has ordinary rotational and other symmetries, but not reflection symmetry. It seems like any vector theory would have to have a reflection symmetry. The standard model is a relativistic spinor field theory, and spinor fields can be left-handed or right-handed.

Many of the molecules critical for life on Earth, such as sugars, amino acids, and DNA, have a right-left asymmetry and only one form is actually used. This is thought to be a quirk of evolution on Earth, and not because of any asymmetry in the laws of physics or chemistry.

The universe is obviously not in a state of symmetry between matter and antimatter. When matter meets antimatter, they both dissolve into pure energy. The universe is made of matter, not antimatter. It would be nice to have some theoretical understanding of why there is vastly more matter than antimatter. The standard model does leave some loose ends like this, but it has been so extraordinarily successful that there are no major anomalies known.

Mathematical physics

Advances in 20th century physics have been led by a misunderstood field called *mathematical physics*. The term does not mean physics that uses mathematics. All physics uses mathematics. Feynman once said, "People who wish to analyze nature without using mathematics must settle for a reduced understanding." Mathematical physics is the study of mathematical proofs that are potentially applicable to physical problems.

Mathematics is fundamentally different from sciences like physics. In physics, the crucial test of an idea is whether some real-world experiment can show that it is superior to the alternatives. Mathematicians accept something as true if it can be rigorously proved by pure logic. They have standards of rigor that are not well appreciated in other fields. Once a mathematician has proved something, there is no need to have any other evidence for it.

A good example is the Pythagorean theorem. It can be proved from the axioms of Euclidean geometry, and no one who understands the proof can have any doubt about its mathematical truth. You can also verify it by measuring the sides of physical triangles, but those measurements do not prove or disprove the mathematical theorem. First, measurements have experimental error, and will not be exactly correct. Second, we know now from general relativity that space is curved and not Euclidean, so the Pythagorean theorem will not be physically true in a gravitational field.

In the 1800s, Gauss developed a mathematical theory of curved surfaces, and Riemann extended it to a theory for higher dimensions. They gave proofs for their theorems just like Euclidean geometry. Perhaps they imagined that the space we inhabit might be curved, but they could not have foreseen that their theory would be the basis for understanding gravity in our 4-dimensional spacetime.

When mathematicians prove theorems about flat (Euclidean) space or theorems about curved space, those theorems are true regardless of whether physical space is flat or curved. Mathematicians find logical truths, not experimental truths. Einstein eschewed mathematical proofs, and explained it this way:

> As far as the laws of mathematics refer to reality, they are not certain; and as far as they are certain, they do not refer to reality.[296]

Mathematicians are like Plato, and believe in a mathematical reality that is the highest form of truth. Physicists are like Aristotle, and always want conclusions about physical reality, based on observations.

A few years before Poincare revealed the relativistic symmetries, he described the purpose of mathematical physics:

> To summarize, the end of mathematical physics is not merely to facilitate the numerical calculation of certain constants or the integration of certain differential equations. It is more, it is above all to disclose to the physicist the concealed harmonies of things by furnishing him with a new point of view.[297]

The French mathematician Alain Connes explains the difference between mathematicians and physicists by saying that mathematicians are like fermions, and physicists are like bosons.[298] Recall that fermions are to bosons as matter is to energy. Fermions have a spin that prevents two particles from occupying the same quantum state, so fermions can combine to form a material substance like a solid body, while bosons can coalesce like a laser beam to exert a force. Physicists like to accumulate similar evidence in order to strengthen their beliefs in a hypothesis, while mathematicians just need one proof. Mathematicians are building a body of knowledge in which each proof is separable from every other proof, and the proofs combine to form the surest form of truth. Physicists join fads and promote the popularity of their work, and mathematicians do the opposite. Poincare was a fermion, and Einstein was a boson.

Mathematicians have just one standard for what a proof is, and they are usually not impressed by physical arguments that cannot be formalized into proofs. As Poincare said:

> How can a demonstration, that is not rigorous enough for the analyst, be sufficient for the physicist? It seems that there cannot be two rigors, that rigor is or is not, and that where it is not, argument cannot exist.[299]

This may explain why Poincare said so little about Einstein. Einstein's arguments were not rigorous, and so they did not exist as arguments that Poincare would recognize.

Mathematicians and their tools have been vital to understanding reality. Some who have contributed crucially to our understanding of early 20th century physics include Poincare, Cartan, Weyl, Hilbert,

Noether, John von Neumann, and Dyson. They probably all would have received Nobel prizes if they had been considered physicists, but the prize is not given to mathematicians.

A big breakthrough in quantum field theory was when the Dutch mathematical physicist Gerard 'tHooft proved in 1971 that gauge theories were renormalizable, and hence fully relativistic. He and a colleague did get the Nobel Prize in 1999 in a rare theoretical prize. Steven Weinberg and others got the prize in 1979 for applying gauge theory to weak forces.

Penrose proved that under general relativity and certain physical assumptions, a black hole must have a singularity at the center, where the density is infinite and spacetime becomes meaningless. Hawking used similar methods to prove that the whole universe must have begun with a singularity. These theorems convinced physicists that black holes and the big bang were essential parts of general relativity theory.

When the Nobel committee gave a prize for the big bang, it went to a couple of telephone company scientists who were just measuring the static that seemed to be interfering with their satellite communications and TV reception. They did not realize it, but they had measured the cosmic background radiation (CMB) that cosmologists had predicted as a residue from the big bang. No prize went to the theorists who actually did the work to understand the big bang. The prize is primarily for important experimental results, not theory.

These are just some examples of sophisticated mathematics being applied to physics. The theories behind symmetry groups and curved space are just two of many examples.

5. Philosophers look for revolutions

Physicists usually ignore philosophers and whatever they have to say. The great German philosopher Immanuel Kant had grand and seemingly important things to say about space and time back in 1787, but it was incomprehensible and irrelevant to physicists. Nobody can agree as to whether relativity proved him correct or proved him wrong.

In the 20th century the physicists became incomprehensible, and the philosophers started describing science in terms of revolutions and paradigms.

The Copernican revolution

According to many historians, modern science began in 1543. That is the most commonly given date for the birth of the so-called *scientific revolution*. The Polish-German astronomer Nicolaus Copernicus published his great book that year. It became one of the more famous books printed by the printing press, which had just been invented in the previous century. About 1,000 copies of the first and second editions were printed, and 600 of them remain today. Only the Gutenberg Bible is more famous.[300]

The scientific revolution brought us the scientific method, and all the luxuries of modern science. The scientific method consisted of making observations, formulating hypotheses, and doing experiments to test those hypotheses. Ancient scholars were not sufficiently enlightened to actually test what was written in their great books. Or so the story goes.

The trouble with this story is that De Revolutionibus was not a particularly good example of the scientific method. While Copernicus had a brilliant model for the solar system, he had no way of demonstrating that it was any better than the prevailing models. The later work by Tycho and Kepler was much more scientific because they had testable hypotheses.

The book title used the word "revolution" in the sense of the orbs revolving around the Sun, not in the sense of an intellectual revolt. That is, a Copernican revolution was just a model for planetary motion. The book had Europe's first detailed heliocentric model of the solar system since Ptolemy's geocentric model more than a millennium earlier.

The book was not particularly controversial. It was published with an endorsement from the Catholic Church. It had been known since ancient Greek times that the solar system could be modeled with a geocentric or heliocentric model, and medieval astronomers did not necessarily attach any physical significance to the choice. An astronomer might even use a geocentric model for one planet's motion while using a heliocentric model for another planet. Such mixed models had been used both before and after Copernicus, and were known to him.

In the preface, Copernicus says that he delayed publication because he feared the scorn of the public. Yet he himself has scorn for those "completely ignorant of the subject", and says "Astronomy is written for astronomers." He says that he describes the motions of the Earth and the general structure of the universe.

Copernicus gave arguments for the motion of the Earth. He justified putting the Sun at the center as being like putting a lamp in the middle of a temple:

> At rest, however, in the middle of everything is the sun. For in this most beautiful temple, who would place this lamp in another or better position than that from which it can light up the whole thing at the same time? For, the sun is not inappropriately called by some people the lantern of the universe, its mind by others, and its ruler by still others. ... the sun governs the family of planets revolving around it.[301]

The book had many pages of geometrical constructions, and tables of figures that had been compiled from ancient observations. It was the first complete new model since the Almagest. There was an anonymous forward that correctly argued that the book could be used for reliable computation, whether the hypotheses about the causes of motion were true or not.

That anonymous forward, written by the German Lutheran theologian Andreas Osiander, has been called the greatest scandal in the history of science, because it suggested that astronomical models could be just useful mathematical fictions for the convenience of astronomers.[302] Some even say that Copernicus felt so betrayed and repulsed by it that he dropped dead upon first seeing his published book. In fact, Osiander's forward is completely reasonable. It defends the work against those who might be offended by its novel hypotheses. It explains that astronomers may find different causes for observed motions, and choose whatever is easier to grasp. As long as a hypothesis allows reliable computation, it does not have to be the truth in some philosophical sense. Astronomical models need not necessarily describe the true causes for motion, but may still be useful as mathematical devices for calculating the movement of planets. The forward concisely explains all these things. There is nothing scandalous about a publisher inserting a forward written by someone else in order to broaden the appeal of the book.

Osiander's view was similar to the Nobel Prize committee's citation to Gell-Mann. That citation suggested that the quark hypothesis allows reliable computation about nuclear interactions, whether or not the quarks exist in some philosophical sense. It is a very modern view because modern physics is filled with useful hypotheses that are subject to different interpretations.

Here is a typical explanation of the Ptolemaic system, and how it was replaced:

> In its final form, the model was extremely complicated, requiring many nested levels of epicycles, and with even the major orbits offset so that they were no longer truly centered on the Earth. Despite all of this fine tuning, there remained significant discrepancies between the actual positions of the planets and those predicted by the model. Nevertheless, it was the most accurate model available, and it remained the accepted theory for over 13 centuries, before it was finally replaced by the model of Copernicus.[303]

Most of this is false. Ptolemy's system did not have nested levels of epicycles. It did have orbits offset from the center, but so did every subsequent model. The model of Copernicus had more epicycles, more fine-tuning, and more discrepancies, and did not replace

Ptolemy's. Copernicus even had epicycles on epicycles, which Ptolemy never had.

Ptolemy's universe is commonly drawn as nested orbits centered about the Earth, but no such diagram appears in the Almagest. That book just used one orb to represent the image of what people see in the sky.

The epicycle became a metaphor for bad science. When a scientist patches a bad theory by making artificial non-physical adjustments to make it conform to observations, he is accused of adding epicycles. But it is not so well-known that it was Copernicus, not Ptolemy, who added epicycles to compensate for faulty physics.

As an example of the epicycle metaphor, the most talked-about evolutionary biology paper of 2010 said that the prevailing theory had Darwinian epicycles.[304] The lead author criticized a rival theory by saying, "It's precisely like an ancient epicycle in the solar system. The world is much simpler without it."[305] But the epicycle was not some sort of unnecessary complication to Ptolemaic astronomy. Saying that the world is simpler without epicycles is like saying that the atmosphere is simpler without clouds. The epicycle corresponded to the revolution of the Earth or the other planet. There is no simpler system for describing that motion, other than using an epicycle or something equivalent to represent the Earth's orbit. If you think that the epicycle is superfluous, just ask yourself how else you would describe the Moon's orbit.

Centuries later, the Copernican revolution became a metaphor for looking at a system from a more objective point of view. The idea is that looking at something from your own point of view is like looking at the night sky from Earth. The Copernican view is more like the view from someone on the Sun, and is not biased toward any particular planet.

In the 20th century, the Copernican revolution took on another meaning, as philosophers settled on Copernicus as defining modern science and as undermining man's role in the universe.

The argument is often made that Ptolemy was unscientific and wrong; that his theory was ad hoc and not real; that he had to keep patching his theory by adding epicycles and using epicycles on epicycles; and that the whole theory had become overly complex and cumbersome.

Copernicus simplified everything by throwing away the epicycles and redefining the center of the universe. The argument is entirely false.

Simon Singh wrote, in a popular cosmology book:

> The Ptolemaic Earth-centered model of the universe was constructed to comply with the beliefs that everything revolves around the Earth and that all celestial objects follow circular paths. This resulted in a horribly complex model, replete with epicycles, heaped upon deferents, upon equants, upon eccentrics. In The Sleepwalkers, Arthur Koestler's history of early astronomy, the Ptolemaic model is described as 'the product of tired philosophy and decadent science'. [306]

No, it wasn't constructed to comply with those beliefs, it wasn't any more complex than necessary, and it certainly was not decadent science. Others have said that Ptolemy required 240 epicycles, or 16 levels of epicycles. But these are absurd exaggerations that were apparently cooked up to justify the claim that the Copernican Revolution was simpler.

Ptolemy's Almagest had epicycles, but only epicycles with direct physical significance. Each epicycle corresponded to a circular planetary orbit. It was Copernicus who added numerous epicycles in order to patch deficiencies of his model, and he had more epicycles than Ptolemy. Copernicus's model was not any simpler either. If anyone, it was Copernicus who had added ad hoc epicycles.

Singh goes on:

> Occam's razor favoured the Copernican model (one circle per planet) over the Ptolemaic model (one epicycle, deferent, equant and eccentric per planet), but Occam's razor is only decisive if two theories are equally successful, and in the sixteenth century the Ptolemaic model was clearly stronger in several ways; most notably it made more accurate predictions of planetary positions.[307]

But the Copernican model was not simpler. It had many circles per planet. It is awkward to give a precise number because it gave multiple methods for calculating orbits, with some using more circles than others. Copernicus also used an eccentric, as the Sun was not truly at the center.

Singh blames the medieval Arabs for having "never doubted Ptolemy's Earth-centered universe with its planetary orbits defined by circles within circles within circles." Later he criticizes another theory for being ad hoc by comparing it to how "Ptolemy had fiddled with the epicycles to match the retrograde motion of Mars."[308] But Mars did not have multiple epicycles or circles within circles. It just had one circle for its orbit, plus another circle to get the view from Earth.

Copernicus thought that he was respecting rotational symmetry by getting rid of the equant, but he was doing the opposite. It is because of rotational symmetry that the planets move with non-uniform speed, because they are conserving angular momentum, and conservation of angular momentum is the logical consequence of rotational symmetry. Copernicus had to add epicycles to compensate for the resulting inaccuracies. Even with extra epicycles, Copernicus never got a system that more accurately predicted the night sky.

Singh criticizes another theory by saying:

> Indeed, such ad hoc tinkering was indicative of the sort of blinkered logic that earlier resulted in Ptolemy adding yet more circles to his flawed epicyclic view of an Earth-centered universe.[309]

Singh has some very good explanations of what science is, and credits the ancient Greeks with being brilliant scientists. He correctly explains that there were some legitimate reasons for believing in an Earth-centered system. In another publication, he got into legal trouble for accusing chiropractors of some unscientific treatments.[310] But he has fallen for the myth that Ptolemy's epicycles were unscientific, and the myth of epicycles on epicycles.

Kepler's system was a great advance over the Almagest, but it was not any simpler to apply. He had the functional equivalent of those epicycles, deferents, equants, and eccentrics. Kepler's system is described in terms of his three laws. His first law says that a planet's orbit is an ellipse, with the Sun at one focus. An ellipse is an oval shape that had been known to the ancient Greeks, and is defined by two focus points on opposite sides of its center. So the Sun is not at the center of the orbit, and the other focus is empty. The second law says that a planet sweeps equal areas in equal times. One side of the orbit is closer to the Sun than the other side, and this law says how the planet goes faster when it is closer to the Sun. The third law gives a formula for the time to complete one orbit, based on the distance from the Sun.

Earth completes one orbit in one year, and the outer planets take longer.

The Almagest eccentric is a point off the center of the orbit, just like Kepler's focus. The equant is just like the empty focus, and controls how the planet speeds up and slows down during its orbit. Whereas Kepler calculates a speed-up when the planet is near an off-center point on one side, Ptolemy calculates a slow-down when the planet is near an off-center point on the other side. To get the view of a planet from Earth, Kepler must subtract the Earth's orbit from the other planet's, and that computation is very much like Ptolemy's deferent-epicycle computation. The Almagest treats the planets as unrelated, so it has nothing like Kepler's third law. Kepler was able to achieve much greater accuracy than the Almagest and De Revolutionibus, but his system was not really any simpler.

Somehow Copernicus and Einstein have become the models of scientific progress. But there are much better examples. Truly great scientific progress comes from a mixture of observation and hypothesis, along with some experimental demonstration of the validity of a hypothesis, and ultimately a new theory that is broadly applicable.

Tycho and Kepler are examples of science at its finest. Every great civilization, from the Babylonians to the Mayans, had astronomers observing and cataloging the celestial movements, and Tycho outdid them all, without even a telescope. He devised a new model of the solar system that finally displaced the Almagest. Kepler outdid Tycho, and devised an even better model that achieved an accuracy that no one thought possible. He turned heliocentrism into a real scientific theory with testable hypotheses. He predicted the phases of Venus, which Galileo soon observed with his telescope. He invented a telescope that was an improvement on Galileo's designs. He suggested that Galileo try to measure stellar parallax, although the telescopes were not good enough until a couple of centuries later. He didn't just propose distances to the planets; he found independent ways of measuring those distances. Their work was later superseded by Hooke-Newton gravitation and by general relativity, but Newton would have been helpless without Tycho and Kepler.

These scientists had some beliefs that are out of favor today, and they are sometimes ridiculed for their beliefs and for being colorful characters. Tycho believed that the Earth was stationary. He had a gold nose because he lost his real one in a duel. Kepler had some mystical beliefs in the harmony of the universe, and wrote the first science fiction novel about a trip to the Moon with some demons. His mother got put on trial for witchcraft, and he had to legally defend her, and to footnote his novel with explanations that it was an allegory. He was excommunicated from his Lutheran church.

Whatever personal shortcomings a scientist has, they are irrelevant. His work speaks for itself. Tycho is remembered for his data, and Kepler for his three laws. Anyone else could have verified what they did by reproducing their work. Well, not anyone, because they had exceptional tools and skills. But they published their works, and what they said was objectively verifiable. They could predict the location of Mars more accurately than anyone else. They are the examples of great science.

Defining the scientific method

Not everyone agrees on what defines great science. Kansas was the center of a hot political debate in 2005 over what had previously been an obscure philosophical issue. The state was revising its educational standards, and the New York Times reported:

> Perhaps the most significant shift would be in the very definition of science - instead of "seeking natural explanations for what we observe around us," the new standards would describe it as a "continuing investigation that uses observation, hypothesis testing, measurement, experimentation, logical argument and theory building to lead to more adequate explanations of natural phenomena."

Philosophers call this issue (of defining science) the *demarcation problem*. Sometimes it is difficult to say what is or is not science. Scientific activists were outraged at this change. The American Association for the Advancement of Science (AAAS) denounced it as a weakening in its support for evolution and cosmology. After a national campaign against school officials, the Kansas definition was changed back in 2007 to:

> Science is a human activity of systematically seeking natural explanations for what we observe in the world around us.

In other words, science is whatever scientists do, as long as they do not invoke God or any other supernatural force. The AAAS did not want to be restricted to testable hypotheses. Teachers can teach unsupported ideas, as long as they are atheistic.

The attack on Kansas was led by Harvard evolutionist Stephen Jay Gould. He wrote in 1999 that Kansas should teach evolution as a fact, "as strongly as the earth's revolution around the sun rather than vice versa."[311] He was elected president of the AAAS in 2000.

In its eagerness to define evolution to be scientific in 1995, the National Association of Biology Teachers declared it to be "an unsupervised, impersonal, unpredictable and natural process". They seemed to be saying that God does not supervise the process, and that science cannot make any testable predictions about it either. They sound like anti-religious mystics, not scientists. They have since backed off this definition, but it continues to cause controversy.[312]

If these definitions are confusing, do not worry. Most scientists pay no attention to such philosophical issues anyway.

The essence of the scientific method is so simple that it is taught to grade school students doing science projects. It means doing observations, making testable hypotheses, and then doing experiments to test those hypotheses. The details vary, but there has to be some way of making statements that might be true or false, and some way of doing observations or experiments on the natural world in order to determine the truth or falsity of those statements.

The scientific method is older than recorded history. When prehistoric cavemen made stone tools, they presumably did systematic experiments to see which rocks would work best. Ancient civilizations built ships and pyramids, invented beer and bread, and accomplished many other great things. The Babylonians measured the period of the Moon with 6-digit accuracy.[313] Their use of the scientific method was extremely impressive.

And yet there is a common myth that the scientific method was invented in the last 500 years. The myth is absurd. What did happen was that the printing press was invented about 500 years ago. Distribution of printed books vastly accelerated the progress of science, because scientists could more easily use the observations of others, and test the hypotheses of others. Knowledge could accumulate much more readily.

This myth goes along with the myth that Columbus discovered that the Earth was round in 1492, as stupid European Bible-readers supposedly believed in a flat Earth. These myths are so silly that it is a wonder that they persist. Columbus did not sail around the Earth. He tried to sail to India and the experts of the day predicted that he would never get there because he did not have nearly enough food and supplies to traverse such a great distance.

The Columbus myth gets cited by those who want to make fun of medieval authorities as being unscientific and foolishly holding on to old ideas until a daring revolutionary explorer proves them wrong. But the king's experts were actually right. Columbus had his own misguided theory that the circumference of the Earth was only about a third of what the experts said. Columbus was wrong and he was lucky to have found other land instead. He boldly led a successful expedition, but he cannot be credited with disproving the flat Earth theory.

Copernicus promoted his own version of the flat Earth myth in his book. He ridiculed his non-astronomer critics by comparing them to an early Christian writer who believed in a flat Earth in about AD 300. Copernicus implies that no astronomer would believe anything so silly.

Aristotle and the ancient Greeks knew that the Earth was round, and they had solid scientific reasons for thinking so. They noticed that lunar eclipses occurred when the Moon was full. They correctly deduced that the Sun was lighting up the Moon when it was on the opposite side of the Earth, and the Earth's shadow was causing the eclipse. The shadow was round. The Greeks observed ships going over the horizon at sea, as if the Earth were round. When they sailed a ship towards land, they saw the mountaintops before the beach. They noticed that the Sun was more directly overhead at midday if they traveled further south. With some clever geometry, they could figure out their latitude by estimating the angle of the Sun at midday. The Sun is directly overhead at the equator on the equinox at noon, and the angle at other places indicates the lattitude. By measuring the distance between lines of latitude, they estimated the circumference of the Earth to within 10%.

Aristotle and Ptolemy thought that it was unlikely that all of the Earth's land is concentrated in one hemisphere. The word *antipodes* referred to a hypothetical continent on the other side. Early Catholic Church scholars debated whether there could be antipodean people.

This Greek analysis was scientific because it was using observations to formulate and test hypotheses. The hypothesis was that the Earth was round and had a circumference of about 25,000 miles. From that and a little Euclidean geometry, they could predict how far out at sea a ship would disappear over the horizon, based on the size of the ship and the height of the observer on shore. They could predict north-south distances by observing the elevation of the Sun at midday. And they generated the knowledge that allowed others to predict how long it would take Columbus to get to India.

All of this was well-understood for two millennia. The flat Earth myth was created in the 1800s, and promoted primarily by Darwinists to ridicule Christians.[314] The reasoning was that if Christians were so wrong about the flat Earth, then they could be wrong about Darwinian evolution also. The myth has continued, as the forward to a recent book by a history professor explains:

> At the beginning of his book he quotes from current text-books used in American grade schools, high schools and colleges which insist that there was a consensus among medieval scholars from A.D. 300 to 1492 that the earth was flat. This also was the thesis of the influential historian Daniel Boorstin writing for a popular audience in his book, The Discoverers, published in 1983. Russell then uses his deep knowledge of medieval intellectual history to demonstrate that the opposite was true. It was conventional wisdom among both early-and late-medieval thinkers that the world was round.[315]

A related myth is that primitive people believed that the Earth was held up by elephants or turtles. Sometimes an elephant is on top of a turtle, and sometimes it is turtles all the way down. Hawking starts his 1988 best-selling book with the turtle myth. It has been told for centuries, and almost always to make fun of Hindus or pagans.[316] There is no record of anyone actually believing that these animals supported the Earth.

The myth that modern science started with the astronomy of Copernicus or Galileo is just as nonsensical as the flat Earth myth or the turtles myth. They were not even particularly scientific in their arguments about the motion of the Earth. Copernicus had an interesting hypothesis about what he called the *triple motion of the Earth*, but no way to say whether it was true or false. He had a model of the universe, but no real way to show whether it was any better or worse than Ptolemy's. He had no substantial new observations of his own, and no significant quantitative advantages. What he had was more of a mathematical reformulation of Ptolemaic astronomy than a great scientific advance.

A history professor, in an essay attacking the influence of religion on former President of the United States George W. Bush, wrote:

> By that logic, teaching flat-earthism, or the Ptolemaic system alongside the Copernican system, is a defense of "free speech."[317]

These have become the canonical examples of false ideas. A quiz about the motion of the Earth commonly gets used as a test of scientific literacy. For example, a recent poll reported:

> Probing a more universal measure of knowledge, Gallup also asked the following basic science question, which has been used to indicate the level of public knowledge in two European countries in recent years: "As far as you know, does the earth revolve around the sun or does the sun revolve around the earth?" In the new poll, about four out of five Americans (79%) correctly respond that the earth revolves around the sun, while 18% say it is the other way around.[318]

The American government uses a similar question as one of eight indicators of public science literacy.[319] In fact, both astronomical views are valid because motion is relative and the law of gravity is written in covariant equations. That has been the consensus of our best astrophysicists for a century. It would be a better test of science to ask some questions about objective truths that can be demonstrated with experiments, rather than to ask about some conventions chosen by textbook writers.

Planck ridiculed relativity skepticism by comparison to flat-earthism. In published 1909 lectures at Columbia University, he said:

> Accordingly, the principle of relativity simply teaches that there is in the four dimensional system of space and time no special characteristic direction, and any doubts concerning the general validity of the principle are of exactly the same kind as those

> concerning the existence of the antipodeans upon the other side of the earth.[320]

Galileo did not invent science, but he did do some interesting science. For example, he measured the acceleration of gravity and found that it was constant (on the surface of the Earth), as had been claimed in textbooks.[321] His clocks were not accurate enough to measure falling objects, so he had the ingenious idea observing objects going down an inclined plane. That slows down an object enough for him to measure its acceleration, and with a little geometry he could deduce the acceleration of a free-falling object. He achieved the best pre-Newtonian understanding of motion.

Galileo also acquired a telescope, made improved telescopes, and made important and striking astronomical observations. But his observations did not really advance his hypotheses in a scientific manner. His contemporaries, Tycho and Kepler, were much better examples of scientists doing astronomy. They made more accurate observations, formulated hypotheses and models, and put it all together to predict orbits more accurately than ever before. Galileo never accepted their theories, in spite of overwhelming evidence.

The best story about how Galileo discovered experimental science is the story about how he refuted Aristotle by dropping a cannonball off the Leaning Tower of Pisa. Galileo's 1638 book *Two New Sciences* is written in dialogue form, and the lead character says:

> Salviati: ... Aristotle says that "an iron ball of one hundred pounds falling from a height of one hundred cubits reaches the ground before a one-pound ball has fallen a single cubit." I say that they arrive at the same time. You find, on making the experiment, ...

The story says that for two millennia everyone just blindly accepted this teaching of Aristotle, and it never occurred to anyone to drop two balls to see if they really did what Aristotle said. Because if they did, they would see that Aristotle's physics was obviously wrong, and they would reject his authority. Only Galileo discovered the scientific method and did the experiment at the Leaning Tower. A recent New York Times science article began with a version of this story:

> Galileo's rolling of spheres down an inclined plane four centuries ago disproved Aristotle's notion that falling (or rolling) ob-

> jects move at a constant speed. That was one of the earliest examples of using experiments to devise and test hypotheses to explain observations.[322]

Almost none of this Aristotle-Galileo myth is true. Aristotle never said anything about falling balls. His work included many scientific observations, experiments, and explanations. His physics was rather primitive as it was not quantitative. He is more famous for what he did in other fields. People did not accept Aristotle so blindly. Galileo never dropped balls from the Leaning Tower. And heavier balls do fall faster when air resistance is significant.

It is extremely silly to claim that Galileo invented the hypothesis test. No one could truly believe that the ancient Egyptians built those pyramids without ever testing a hypothesis. Hawking's popular 1988 book heaped even greater, and more absurd, praise on Galileo:

> Galileo, perhaps more than any other single person, was responsible for the birth of modern science. His renowned conflict with the Catholic Church was central to his philosophy, for Galileo was one of the first to argue that man could hope to understand how the world works, and, moreover, that we could do this by observing the real world.[323]

The Kansas controversy was over the teaching of evolution in the public high schools. The evolutionists wanted to make sure that everything in evolutionary biology is regarded as scientific, and that science teachers are not allowed to teach anything that is inspired by a religious viewpoint. So they prefer to take a very broad view of what science is, as long as it is restricted to natural explanations, and not religious ones.

The evolution critics wanted to qualify the definition of science with: "These explanations ought to be testable, repeatable, falsifiable, open to criticism and not based upon authority." They argued that certain aspects of evolution theory did not meet this standard, and should not be regarded as scientific. There were also those with religious views who resented the legal doctrine that forbade teaching anything that was motivated by religious considerations.

The philosophers who have debated the demarcation problem have focused on physics as the prototypical science. Physics has the sharpest division between what is true and what is false. It is governed by mathematical formulas and laws that make precise predictions when given precise data. These laws can be demonstrated with quantitative laboratory experiments. That is why physics is called a *hard* science. The term does not mean that physics is mentally difficult, but that it is firm like a rock. The theory can match the data to five or more decimal places. At the other extreme are *soft* sciences like psychology, where experimental error of only 10% is considered terrific, and where the conclusions are more qualitative than quantitative. Biology is a mixture, with some sub-fields like biochemistry and genomics being hard sciences. The hard sciences are able to measure their success quantitatively while the soft sciences are more qualitative. Fields of study that pretend to be scientific but fail to meet scientific standards are called *pseudoscience*.

It is very much easier to demarcate science in the hard sciences. If some theory makes a non-obvious quantitative prediction that can be verified in a laboratory experiment, then it is a good bet that everyone will consider it scientific. The answers are less clear-cut in the soft sciences. Sigmund Freud became world-famous a century ago, and is still hailed by some as the father of psychiatry and psychology and one of the greatest and most influential scientific geniuses of the 20th century. He is cited more than Einstein and Darwin combined.[324] And yet others have always attacked him as being completely unscientific. There is no firm scientific corroboration for any of his theories. One of his critics, the Austrian and British philosopher Karl Popper, said that his theories are not falsifiable, and thus cannot be scientifically tested. According to some critics, Freud even faked much of his work.[325] And yet his popularity continues because his theory does not make the sort of hard (i.e., firm) predictions that are easily falsified.

The Catholic Church would have had a much harder time arguing with Kepler than Galileo, because he had a quantitative model that outperformed the alternatives. Galileo's arguments were much softer, and the scientific merits of those arguments were more debatable.

The evolutionists eventually persuaded Kansas to purge the word "falsifiable" from its education standards. Many modern philosophers have also backed off the concept. Even many theoretical physicists have rejected the concept, because it undermines some untestable ideas that they like.

There are now prominent physicists who promote the *anthropic principle*, and would like it to be considered science. That principle says that there are some very remarkable coincidences that make human life on Earth possible, but those coincidences are not really coincidences because if there were no human life then we would never notice. There is no laboratory experiment to test this principle.

These physicists get annoyed if you accuse them of doing pseudoscience. They would much rather have science defined in terms of being a human activity. If people look and talk and act like scientists, then they are doing science, some say. At a 2011 physics debate, the American theoretical physicist Lee Smolin gave this definition:

> Science is not about what's true, or what might be true. Science is about what people with originally diverse viewpoints can be forced to believe by the weight of public evidence.[326]

The trouble is that ideas like the anthropic principle are similar to the religiously-motivated concepts that the AAAS sought to eliminate in Kansas. The biggest difference is that the religious folks can be seen attending church on Sunday.

Falsifiability remains the best simple test for what is or is not science. The anthropic principle is philosophy, not science. There is no way to disprove 2+2=4, so that equation is mathematics, not science. An example of science might be the claim that there is an undetectable aether. That could be falsified by detecting the aether somehow. It would not be falsified by formulating another theory that there is no aether, as there is no observation that could ever distinguish between an undetectable aether and no aether at all.

As an example of a scientific hypothesis, consider the claim that a meteor wiped out the dinosaurs. Some imaginative person could have made such a claim a long time ago, but it would not have been considered particularly scientific unless he proposed a way to test it. The claim became a scientific theory in 1980 when it was linked to several testable hypotheses. Namely, it was claimed that a 65 million year old layer of rock would have an unusually high concentration of iridium throughout the Earth, and that all of the dinosaur fossils would be below or in the layer.

Verifying the hypotheses does not quite prove the dinosaur-meteor theory, because there is also a theory that prolonged eruptions from volcanoes (in what is now India) wiped out the dinosaurs. The volcano theory also has its testable hypotheses. Either of these theories might be falsified, and then people would probably believe the other theory. It may not be possible to prove one of these theories to be correct. There could be other possible extinction causes that might not have left any obvious traces. If so, a cautious scientist will not come to any definite conclusions. Science is all about analyzing hypotheses that are testable, not speculating about those that are not.

Science is not just about an accumulation of knowledge about the natural world. Science is also about knowing the limits to that knowledge, and about knowing our ignorance. The best scientist is not the one who first jumps to a conclusion on some issue like global warming, and that conclusion is later proved correct. The best scientist is the one who correctly explains what is deducible from the available data, and correctly understands the limitations of the available knowledge. Ancient Greek astronomers understood that multiple hypotheses could sometimes be consistent with the available data, and the same is true today.

Knocking man off the pedestal

The science historian and evolutionist popularizer Stephen Jay Gould used to love to quote Freud, and his favorite story this:

> Sigmund Freud often remarked that great revolutions in the history of science have but one common, and ironic, feature: they knock human arrogance off one pedestal after another of our previous conviction about our own self-importance. In Freud's three examples, Copernicus moved our home from center to periphery, Darwin then relegated us to 'descent from an animal world'; and, finally (in one of the least modest statements of intellectual history), Freud himself discovered the unconscious and exploded the myth of a fully rational mind.

Gould had a Marxist view of history that exaggerates the importance of revolutions, but he and Freud weren't the only ones to make this point. The biochemist James D. Watson wrote a recent book on DNA, and he starts by giving his version of the three great revolutions. He says that they were the Copernican revolution, Darwin showing that man is a modified monkey, and the Watson-Crick discovery of the molecular structure of DNA!

No, the Copernican revolution was not such a great advance. And neither was Freud's unconscious.[327]

New Scientist magazine recently consulted a panel of experts to determine who did the most to knock man off his pedestal, Galileo or Darwin. The majority favored Darwin.[328]

Current attempts to knock man off the pedestal include theorizing about alternate universes and searching for intelligent life on other planets. Astronomers have recently found many planets orbiting nearby stars, but nothing habitable. It took three billion years for life on Earth to evolve from one-celled organisms to two-celled organisms. For this to happen, the Earth had to be geologically alive and cosmologically stable over a long time. The stability comes from, among other things, having a single large Moon, and having at least one large outer planet (Jupiter). No one has found a similar configuration elsewhere.

The Columbia University physicist and string theory popularizer Brian Greene argues that mathematical reasoning has led to a sequence of "cosmic demotions" that have knocked us out of the center, and following the Copernican pattern may lead us to conclude that our universe is not at the center of the multiverse.[329] He finds it "parochial" to limit science to what can be substantiated with observational evidence.

Some modern writers attempt to draw great philosophical significance from the Sun being at the center of Copernicus's system, and the Earth being at the center of Ptolemy's system. But those statements are misleading. Copernicus put the Sun near the center, not at the center. Most of those who used these systems did not attach great philosophical significance to being at the center. The idea that astronomers might have preferred the Ptolemaic theory so that man would be at the center of the universe was cooked up centuries later. It was not a major concern of the ancient or medieval astronomers. There is no record of anyone being upset about man not being at the center. They were more likely to think that Hell ought to be at the center.

It is a little misleading to say that Ptolemy put the Earth at the center of the universe. He did not really put the Earth at the center of everything. The main reason his system is said to have the Earth at the center is in his description of the daily rotation. Every day the Sun and stars are seen to rise in the East and set in the West, and he understood that either the Earth was rotating or everything was rotating around the Earth. The Greeks figured that out 500 years earlier.[330]

Ptolemy said that a rotating Earth would be simpler, but his model did not make that choice, so the Earth was described as being at the center of the daily rotation. But the Earth was not at the center of the circles used to represent the orbits of the Sun, Moon, and planets.

The myth that ancient astronomers wanted man to be at the center of the universe is widespread. A 2010 Scientific American article referred to "our desire to place ourselves at the center of the universe",[331] as if that were some universally accepted human desire. It was not. Aristotle gave an assortment of physical and metaphysical reasons for believing the Earth to be stationary, but a desire to be at the center was not one of them. Ptolemy might have tried to avoid eccentrics,zz-what? if he wanted the Earth at the center of the universe. Later astronomers in the following centuries did not try to remove eccentrics either. Ancient accounts of the arguments for and against geocentrism are more scientific than most of the modern ones.

Many people go further, and portray all of modern science as a sequence of discoveries about what we don't know, rather than what we do know. It is as if everyone once thought that Aristotle knew everything, and we have been discovering ever since that we know nothing. They say that Einstein showed that everything is relative, quantum mechanics showed that everything is uncertain, and logicians showed that truth is not necessarily provable. The Aristotelian and Newtonian worldviews were demolished.

For Einstein, the aether is the pedestal. The aether was supposedly some sort of crutch that 19th century scientists hung onto because of some misguided view of their own self-importance. By abolishing the aether, Einstein was like Galileo attacking Aristotle or Scripture. Or so the myth goes.

Nearly all of the popular philosophical implications of modern physics are mistaken. Quantum mechanics does have its uncertainty principle, but it also achieves certainties that were previously impossible. Computers and lasers are designed with quantum mechanics, and they are precise and predictable. In Newtonian mechanics, planets can orbit the Sun in lots of ways. In quantum mechanics, electrons can only orbit an atom's nucleus in a few discrete ways.

Newtonian mechanics is deterministic while quantum mechanics is probabilistic, but there is not much practical consequence to this distinction. Poincare showed that a Newtonian gravitational system can become chaotic, meaning that precise long-term predictions are impossible. Modern chaos theory shows that many other systems, Newtonian or otherwise, can similarly become unstable.

Aristotle's physics was not so much overthrown, as disregarded. It did not make any quantitative predictions.

Relativity has been confused with moral relativism, and other doctrines that deny absolutes. The idea is that if Einstein could figure out that there is no absolute space or time, then maybe there are no absolute standards in morality, politics, art, religion, or anything else.

There is some question about whether the analogy is even apt. The essence of relativity is not really that there are no absolutes. All of the formulas and consequences of relativity would be the same whether there are any absolutes or not.

The name "relativity" came from Poincare in his popular 1902 book. Einstein did not even like the term at first, and wrote about the "so-called relativity theory" in 1909. Klein suggested in 1910 calling it "invariance theory", because the essence of the theory is that the laws of physics are invariant under the Poincare group.

As an analogy, you can measure distance in either feet or inches. Whenever you give a numerical figure for a distance, you need to say whether it is relative to a foot or an inch. In that sense, measuring distance is relative. But that does not say anything about whether there is any absolute distance.

To mathematicians, it makes perfect sense to describe the spacetime manifold and metric tensor as being absolute. Manifolds automatically allow different coordinate systems, and formulas on a manifold can always be written in those coordinates. Minkowski even suggested that the relativity postulate be renamed "the postulate of the absolute world", as the theory makes spacetime an absolute concept. Space and time are commensurable by the speed of light, and that is an absolute constant.

The point is that there might be no absolute x-axis because any chosen x-axis could just as well be an y-axis or a z-axis. There are symmetries that relate those axes, and there might be no neutral way to distinguish them. Likewise, there are Lorentz symmetries that relate space and time, so there might be no way to separate them into an absolute space and absolute time. But spacetime together could be an absolute object.

The name relativity is appropriate because the theory teaches that the laws of physics can be written relative to any reference frame. But when you phrase it that way, it does not sound any different from Newtonian physics, as its laws can also be written in any inertial frame. The novelty of relativity is that different frames are related by the Poincare group, as opposed to some other group.

At any rate, relativity does not just mean that everything is relative and that man's importance in the universe is diminished. And it certainly does not mean *moral relativism*, which is the idea that morality is relative to culture and not based on universal truths.

Paradigm shifts

A seemingly obscure book on the history and philosophy of science is credited with being one of the most influential books of the 20th century. Former Vice President of the United States Al Gore said that it was the best book he ever read. The book was *The Structure of Scientific Revolutions* (1962) by Thomas Kuhn, and it introduced the term *paradigm shift* into the lexicon. It has sold over a million copies in 25 languages. At one time, it was the most frequently cited book in the scientific literature. By the year 2000, usage of the word "paradigm" overtook terms that were previously far more popular, such as quantum, symmetry, and chaos.[332]

The book portrayed the history of science in terms of paradigm shifts interrupting normal science. A paradigm is the worldview of the scientists, and a paradigm shift is a revolutionary change in that worldview.[333] These changes do not happen because scientists rationally determine that the new view is superior, but rather because the old view goes out of fashion and its proponents die out.

The changes are compared to religious conversions that defy rational explanation. In the words of one scholar, Kuhn reduced science to "nothing more than long periods of boring conformist activity punctuated by outbreaks of irrational deviance."[334] The book was a direct attack on the idea that scientists are rationally working towards the truth about how the world works. It cast doubt on whether great scientific advances were really any better than the older theories. In short, Kuhn denied that science was progressing towards the truth. He even denied that truth and reality were meaningful concepts.

Most people in the hard sciences do not subscribe to any of this nonsense. They believe that they are studying an objective reality, and that they are making progress doing it. Some science journal editors have a policy forbidding the use of the word "paradigm".

Outside the hard sciences, though, this book is the Bible on how science works. It is accepted uncritically in universities. It is important to understand because it has so thoroughly infected popular opinion on what science is all about. There are many philosophers and other academics who are even more anti-realist than Kuhn. They teach theories that go under names like postmodernism, deconstructionism, and strong program. They deny objective reality as it has been understood since Aristotle.

Kuhn did not say that science was mob rule, but that was the lesson that people learned. Sometimes the followers of a movement are more radical than the chief guru. Kuhn did describe normal science as being done by tradition-bound scientists who grind away within the dominant paradigm and refuse to objectively evaluate a radical new idea. Occasionally a young revolutionary will appear and suddenly change everything. The older scientists are not persuaded, and change only occurs when they are replaced by a younger generation. The bumper sticker slogan "subvert the dominant paradigm" encourages the revolutionaries.

The standard examples of paradigm shifts, Copernican astronomy and Einstein's relativity, do not match this description at all.

Acceptance of heliocentrism was not sudden; it was spread over 18 centuries. Copernicus did build on the previous Ptolemaic system, as he used the epicycles extensively. The heliocentric and geocentric theories were commensurable, as astronomers of the day compared their predictions for planetary orbits. It was Kepler who developed the first really good heliocentric model, but he built on the geocentric models of Ptolemy and Tycho, not the Copernican model.

Special relativity was not invented by a young revolutionary. Poincare was 51 years old in 1905, and he was indoctrinated in the accepted theories, and he built on them. He was reconciling the decades-old Maxwell theory with the centuries-old relativity principle. He believed that they were compatible when everyone else had given up. The others credited with work on relativity were not so young either. While Einstein was 26 years old, Lorentz was 52, Planck was 47, and Minkowski was 41. FitzGerald would have been 54, if he had still been alive. In 1915, when Einstein is credited with his best work on general relativity, he was 36, Grossmann was 37, Levi-Civita was 42, Hilbert was 53, Schwarzschild was 42, de Sitter was 43, and Weyl was 40. All had well-established reputations, except for Einstein in 1905. Relativity differed from Newtonian mechanics most obviously in the prediction that an accelerating electron would show relativistic mass. This difference was certainly commensurable, as Kaufmann was already doing experiments to measure the predicted effect in 1901.

The younger theorists were not any bolder than the older ones either. Poincare applied his principle of relativity to all the laws of physics while Einstein was sticking to electromagnetism. Poincare said that the aether was unobservable while Einstein only said that it was superfluous to his derivation. Poincare was willing to say that other physicists were wrong, while Einstein was not. Poincare and Minkowski were promoting a 4-dimensional spacetime while Einstein had trouble accepting the concept. Poincare presented radical new ideas while Einstein rehashed old ones.

At a 1906 conference where physicists were comparing relativity to the newer alternatives, Sommerfeld said that it was the older physicists who preferred the Lorentz-Einstein theory. He said, "I suspect that the gentlemen under forty will prefer the electrodynamical postulate, while those over forty will prefer the mechanical-relativistic postulate."[335] By then the Lorentz contraction was 17 years old, and was favored by the more senior experts. The newer theory with the "electrodynamical postulate" was the electron theory developed by the 31-year-old German physicist Max Abraham, and Einstein was in the position of defending the older and more established relativistic theory. The two theories disagreed about how much mass should increase with velocity, and the experimental evidence was inconclusive at the time.

Those who say that Poincare was too old and too conservative to accept radical new ideas are uninformed. He made many major original conceptual advances in his special relativity papers. He said that local clocks measure local time and that the electromagnetic relativity principle holds to all orders, although he generously attributed these ideas to Lorentz. He declared the universality of the speed of light, and that the aether is unobservable. He found and named the Lorentz group and its infinitesimal generators (known as the *Lie algebra*, and fundamentally important in quantum mechanics), discovered the spacetime metric, and deduced relativity as a consequence of the geometry of spacetime. He invented imaginary time as a way to explain the symmetries of spacetime. He discovered the simultaneity paradox, and was led to an operational definition of time and space. He invented 4-vectors, and proved the covariance of Maxwell's equations. He formulated a relativistic action, and used it to give another covariance proof. He proposed a Lorentz-invariant theory of gravity, with gravity waves that propagate at the speed of light. There is no record of any independent discovery of these concepts, and all were absorbed into mainstream physics in a way that can be directly traced to Poincare.

More importantly, acceptance of heliocentrism and relativity was nothing like the non-rational cultural shift that Kuhn describes. Scientists of the day had perfectly good reasons for believing what they did, and they readily changed when the hard evidence came in. Even the opinion of the 17th century Pope is defensible today.

Kuhn got many of his ideas from the Hungarian philosopher Michael Polanyi. He had a philosophy of science based on Copernicus and Einstein, and he was also hung up on revolutions of various sorts. He was interested in the separation between reason and experience, and argued that the scientific method was overrated. He said objectivity is a delusion, and he preferred reason. The true lesson of the Copernican revolution is that we prefer the more ambitious anthropocentrism of our reason over that of our senses, according to him. The astronomy of Copernicus and Galileo was not superior to that of Ptolemy by any objective measure, except by being more aesthetically pleasing.

Polanyi draws a similar lesson from the history of relativity, and argues that Einstein was guided by pure reason:

> The usual textbook account of relativity as a theoretical response to the Michelson-Morley experiment is an invention. It is the product of a philosophical prejudice. When Einstein discovered rationality in nature, unaided by any observation that had not been available for at least fifty years before, our positivistic textbooks promptly covered up the scandal by an appropriately embellished account of his discovery. ... To make sure of this, I addressed an inquiry to the late Professor Einstein, who confirmed the fact that "the Michelson-Morley experiment had a negligible effect on the discovery of relativity".[336]

Time magazine reported in 1970 that it was a myth that Einstein followed experiment:

> Yet was Einstein actually guided toward his epochal achievement by the Michelson-Morley experiment? ... Holton concludes that the answer is no. ... Einstein speaks of the influence ... in such words as "negligible," "rather indirect" or "not decisive." Furthermore, toward the end of his life, Einstein appears to have become increasingly determined to demolish the myth. In an unpublished letter written only a year before his death, Einstein said: "I even do not remember if I knew of [the experiment] at all when I wrote my first paper on the subject." ... Holton says, textbook writers (himself included) have nurtured what he calls the "experimenticist fallacy": the false notion that theory always flows directly from experiment.[337]

A more recent book says something similar:

> Another puzzling fact about Einstein's paper is that it did not mention the Michelson-Morley experiment or, for that matter, other optical experiments that failed to detect an ether wind and that were routinely discussed in the literature concerning the electrodynamics of moving bodies. There is, however, convincing evidence not only that Einstein was aware of the Michelson-Morley experiment at the time he wrote his paper, but also that the experiment was of no particular importance to him. He did not develop his theory in order to account for an experimental puzzle, but worked from much more general considerations of simplicity and symmetry. These were primarily related to his deep interest in Maxwell's theory and his belief that there could be no difference in principle between the laws of mechanics and those governing electromagnetic phenomena. In Einstein's route to relativity, thought experiments were more important than real experiments.[338]

No, those thought experiments were only given as a way of explaining the work of others, and had no role in the development of the theory. Another scholar praises Einstein's originality and says this of his 1905 paper:

> It looks as though he had reached the conclusions by pure thought, unaided, without listening to the opinions of others. To a surprisingly large extent, that is precisely what he had done.[339]

This conflicts with Einstein's own accounts earlier in his life, when he stressed the importance of Michelson-Morley for relativity. The only reason anyone can say that Einstein ignored the Michelson-Morley experiment is that he restated in postulate form the consequences that Lorentz deduced. By the time that Einstein wrote his first relativity paper, Lorentz's prediction of relativistic mass had already been experimentally tested. More importantly, Polanyi asked the wrong man. Relativity theory was discovered by Lorentz and Poincare, and they were certainly strongly influenced by that experiment. So was Minkowski, as his famous 1908 essay started by saying that the new theory was "grown from the soil of experimental physics." From this mistaken history of relativity, Polanyi gets an entirely wrong view of what science is all about.

The chief difference between Einstein's special relativity and the works of his rivals is that he postulated what they proved with mathematics and experiment. And then 20th century philosophers somehow decided that a postulate is better than a proof and gave Einstein all of the credit.

Some philosophers go further, and argue that relativity might not have been discovered for decades if Einstein had not written that 1905 paper.[340] They say that many other great scientific discoveries were simultaneously made by scientists who were theorizing about the observations of the day, but relativity was different because Einstein ignored all of that and made such a sharp break from all existing physics. This argument is directly contrary to the actual history of relativity, as it was Lorentz and Poincare who made the big breaks, not Einstein, and they built on previous work.

Einstein liked to give the impression that he solely invented special relativity in a flash of brilliance. According to a recent article:

> When asked by the biographer Carl Seelig if a definite birth date could be assigned to the theory of relativity, Einstein wrote back "Between the conception of the idea of the special theory of relativity and the completion of the corresponding published paper there passed five or six weeks" (Seelig 1960, p. 114). …
>
> Similarly, the gestalt psychologist Max Wertheimer reports that Einstein had indicated to him in 1916 that "from the moment, however, that he came to question the customary concept of time, it took him only five weeks to write his paper on relativity

> – although at this time he was doing a full day's work at the Patent Office" (Wertheimer 1945, p. 214).[341]

Just as Einstein promoted the myth that he developed special relativity out of pure reason, he pushed similar myths about general relativity. In 1938, he wrote:

> The deviation of the motion of the planet Mercury from the ellipse was known before the general relativity theory was formulated, and no explanation could be found. On the other hand, general relativity developed without any attention to this special problem. Only later was the conclusion about the rotation of the ellipse in the motion of a planet around the sun drawn from the new gravitational equations.[342]

These myths are in relativity textbooks, such as:

> Nevertheless, it is clear that Einstein was led to GR [general relativity] primarily by his philosophic desire to abolish totally the role of absolute space from physics.[343]

But in fact Poincare had announced in 1908 that the Mercury anomaly could be partially explained by relativity, and Einstein's letters show that finding a fuller explanation was a major purpose in his looking for a relativistic gravity theory.[344] And Einstein's 1915 analysis of Mercury was not even based on his new gravitational equations, but on Grossmann's 1913 equations.

The myth that Einstein revolutionized physics while ignoring everyone else is widespread. It is an inspiration to crackpots everywhere. A book on physics cranks said:

> I would insist that any proposal for a radically new theory in physics, or in any other science, contain a clear explanation of why the precedent science worked, ... the crank is a scientific solipsist who lives in his own little world. He has no understanding nor appreciation of the scientific matrix in which his work is embedded ... In my dealings with cranks, I have discovered that this kind of discussion is of no interest to them.[345]

The cranks are just following Einstein's example, as that example has been described by philosophers who say that he applied pure reason while ignoring prior theories and experiments.

Scientific progress

Before Kuhn, the history of science was usually described in terms of progress. Scientific discoveries and breakthroughs that led to the most progress were emphasized the most. Kuhn turned that idea on its head, and focused on the paradigm shifts that he said did not constitute progress. There was normal science between the paradigm shifts, but Kuhn's view was that the scientific revolutions and paradigm shifts did not bring any measurable improvements to the previous theories.

Kuhn makes an analogy to Darwinian evolution. He argues that Darwin's main innovation was to say that life does not make progress when it evolves. Others had described the history of life on Earth as progress from lower animals like fish to human beings. Gould was influenced by Kuhn and also denied evolutionary progress while emphasizing randomness. Not everyone agrees that evolution lacks progress.

A much better view of science was proposed by the Austrian and British philosopher Karl Popper. He believed that science describes reality, and that experiments help validate scientific theories. He said that scientific theories should be *falsifiable*. That is, scientific theories must have statements or predictions that can be tested by experiment and potentially proved false. Otherwise the theory is not really scientific. Popper developed his philosophy from Poincare's, and described Poincare as the greatest philosopher of science ever.[346]

Popper was inspired to his view of science by relativity.[347] Einstein argued for testing relativity by doing decisive experiments, such as measuring the gravitational deflection of starlight.[348] That made relativity much more scientific than heliocentrism, because Aristarchus and Copernicus never proposed any experiment that would show that heliocentrism was right and geocentrism was wrong. Popper might not have realized that those experiments were not as decisive as the newspapers portrayed, because they had large errors. Good quantitative confirmation of general relativity only came decades later.

Popper even questioned whether it was scientific for Lorentz to redefine space and time to explain the Michelson-Morley experiment. The length contraction had originally been proposed by FitzGerald as an ad hoc way to reconcile the aether theory with the experiment. At the time, Michelson-Morley was the only known second-order aether-wind experiment, and the length contraction had no known observable consequences except to say that the experiment would fail to detect the motion of the Earth. No one could measure the contraction with a meter stick because the meter stick would also be contracted. It seemed like a cheat to redefine space and time whenever an experiment fails to match the theory.

Suppose you had a theory that all lizards were less than a meter long. Then someone reported finding a much larger Komodo dragon in Indonesia, and you explained it away by arguing that it was not really three meters long, but just looks that way because the observer's meter sticks contract whenever they get close to a Komodo dragon. The philosophers would say that you had an ad hoc theory, and everyone else would just laugh at you.

FitzGerald's length contraction hypothesis must have seemed similarly preposterous and untestable at the time, but there was more to it than that. The Lorentz transformations are precisely the symmetries of Maxwell's equations, and there was a lot of evidence for those equations. The original Michelson-Morley experiment required just a length contraction, but similar experiments require transformations of length and time. So the theory was testable, because those were exactly the same transformations as the Lorentz transformations. Many other aspects of the theory, such as the mass increase with velocity, were also tested.

Even if you think that the early evidence for relativity was weak, that was the historical origin of the theory, and no one ever conceived it in any other way. Einstein's 1905 description of the theory was just as ad hoc as Lorentz's. They were both working out the logical consequences to the speed of light being the same for all observers, and modifying the lengths of meter sticks accordingly. The main difference was that Lorentz explicitly relied on Michelson-Morley to derive the light postulate and then the transformations, and Einstein skipped the Michelson-Morley step.

The 2010 Hawking-Mlodinow book describes Einstein's 1905 theory and the previous FitzGerald-Lorentz theory with nearly identical terms. It credits Lorentz and FitzGerald with suggesting that "in a frame that was moving ... clocks would slow down and distances would shrink, so one would still measure light to have the same speed." Ten years later, Einstein assumed that "the speed of light should appear to be the same to all uniformly moving observers", and drew the "startling conclusion that the measurement of the time taken, like the measurement of the distance covered, depends on the observer doing the measuring." But it was Einstein's idea that demanded "a revolution in our concept of space and time." That is how the book describes the difference between Lorentz and Einstein. The book discusses many great ideas in the history of physics, but reserves the word "revolutionary" for heliocentrism (attributed to Aristarchus) and for Einstein's relativity.[349] This book is an example of how physicists have adopted Kuhn's terminology. Einstein's special relativity is presented as only a revolution in the Kuhnian sense of being a restatement of a previous theory, with no measurable or rational advantages. It is only preferred for obscure ideological reasons.

Popper used Freud as an example of unscientific theorizing. Freud had systems for interpreting dreams, for example, but there was no way to tell whether his interpretations were true or false. No matter what the evidence, Freud always seemed to have some ad hoc explanation. When his imagination ran dry, he might say that sometimes a cigar is just a cigar. Popper also attacked the Copenhagen interpretation of quantum mechanics.

It is important to understand that the theory of quantum mechanics is certainly scientific. It makes a great many quantitative predictions that can be experimentally verified. An experiment that gave different quantitative results would falsify the theory. The theory makes potentially falsifiable predictions, but it has not been falsified.

The Copenhagen interpretation is just one of several interpretations for quantum mechanics, and there is no proof that any one of them is any better than any other. The Copenhagen interpretation is like saying that the Moon only exists when someone is looking at it. There is no way to falsify that statement. Under the Copenhagen interpretation, anything you do to observe the Moon will cause it to exist, so there is no way to observe the Moon not existing.

The science crackpots love the Kuhnian paradigm shift philosophy. They can propose some goofy theory, and they don't have to show that it is any better than any other theory. Instead they will just complain that they are ignored for social reasons.

Kuhn's revolutions

There are two main 20th century philosophies of science, Popper's falsification and Kuhn's paradigm shift. The crucial difference lies in what a new theory has to do to replace an old theory. Popper said that the old theory must be shown to be faulty, and the new theory ought to be demonstrably superior. Kuhn argued that such a rational change is impossible, and that new theories are adopted for social reasons. Kuhn's followers like to deny that falsification is a useful concept for scientific practice. Others wonder how anything can be called scientific unless there is some way to verify or falsify it. The associations with Popper and Kuhn are confusing, because Popper denied being a positivist and Kuhn denied being a Kuhnian. Hawking said, "Any sound scientific theory ... should in my opinion be based on the most workable philosophy of science: the positivist approach put forward by Karl Popper and others."[350] Many others follow Kuhn, particularly those outside the hard sciences.

Popper's philosophy is closely related to *positivism*. Positivists believe in the knowledge that can be positively demonstrated by empirical methods. They have great confidence in reducing all valid questions as being objectively answerable by observations and experiments, and rejecting the other questions as meaningless. Poincare thought that all physicists might become positivists some day, if they could resist metaphysical speculations.[351] A positivist sticks to hard science for his beliefs, and tends be skeptical about unobservables.

Today, a great many research papers in the soft sciences try to show that they are scientific by quantifying how well they falsified something. The idea is that they formulate some unlikely hypothesis, and call it the *null hypothesis*. Then they test it with an experiment, and report how the outcome differed from what would be predicted from the null hypothesis. From that they compute a *p-value*, which is the probability of the difference being that large, assuming again the null hypothesis along with an assortment of other hidden assumptions about the fairness of the experiment. If the p-value is less than five percent, then the result is considered *statistically significant*, and hence publishable in a reputable journal. Informally, it is suggested that the chance of the original hypothesis being true is less than five percent, so the scientific community can consider it falsified.

This use of p-values is often criticized.[352] Some say that most medical studies fail to be replicated, and the abuse of p-values is a major cause. Statisticians say that the statistical models are oversimplified and misinterpreted. They say correctly that the p-value does not measure the probability of the null hypothesis being true. But no one wants to take experimental drugs based on a Kuhnian paradigm shift, and so the use of p-values will continue.

Usage of the word "revolution" in the history of science is confusing. It can mean the act of revolving in a circular orbit, as in the Copernican revolution of the Earth about the Sun. It can mean a turning-around, as in looking at a scientific problem from a different point of view. It can mean an intellectual revolt, as in a violent political revolution that overthrows a prior power structure. Or it can mean a large and dramatic advance, as in the misuse of the term quantum leap.

Americans think of the 1776 American Revolution as not just a power reversal, but a great and profound advance in politics, law, and government that ultimately was a very large benefit to the entire Earth. Revolutions in other countries have sometimes been stupid bloody wars that replaced one dictator with another without benefit to the people. A political revolution is not necessarily a change for the better, may not even have much practical consequence to the typical citizen.

Kuhn's favorite example of a revolution was what he called the *Copernican revolution*, and he did not consider it much of an advance at all. He wrote:

> The preface to the De Revolutionibus opens with a forceful indictment of Ptolemaic astronomy for its inaccuracy, complexity, and inconsistency, yet before Copernicus' text closes, it has convicted itself of exactly the same shortcomings. Copernicus' system is neither simpler nor more accurate than Ptolemy's. ... Judged on purely practical grounds, Copernicus' new planetary system was a failure; it was neither more accurate nor significantly simpler than its Ptolemaic predecessors.[353]

Kuhn adamantly maintains that Copernicus's system was no simpler and no more accurate than Ptolemy's. Copernicus did not solve any outstanding problems of the day, and offered no convincing objective reason to prefer his system. As Kuhn tells it, acceptance of Copernicanism was just a function of how broad-minded people were, and he says that medieval scholars were less dogmatic than the ancient Greeks who rejected similar ideas from Aristarchus, whom he describes as having a "complete anticipation".[354] He wrote:

> The answer to this question is not easily disentangled from the technical details that fill the De Revolutionibus, because, as Copernicus himself recognized, the real appeal of sun-centered astronomy was aesthetic rather than pragmatic. To astronomers the initial choice between Copernicus' system and Ptolemy's could only be a matter of taste, and matters of taste are the most difficult of all to define or debate. Yet, as the Copernican Revolution itself indicates, matters of taste are not negligible. ... But only astronomers who valued qualitative neatness far more than quantitative accuracy (and there were a few – Galileo among them) could consider this a convincing argument in the face of the complex system of epicycles and eccentrics elaborated in the De Revolutionibus.[355]

Under paradigm shift theory, the young radical revolutionary has a hard time convincing his dogmatic elders of the merits of his ideas. This is because Kuhn's favorite examples involve a revolution in the sense of a turning-around to a different view of the same thing. When the revolutionary view is not demonstrably superior in any way, it is perfectly natural that some scientists would be unconvinced of the necessity for change.

Kuhn's revolutionaries do not include those who were truly advancing the state of the art. Tycho invented new astronomical instruments, devised a new model that solved the problems of the day, and generally increased astronomical precision by a factor of ten, but he gets little credit. He was just a normal scientist doing normal science, and not a revolutionary. If anything, he was a counter-revolutionary, in Kuhn's view. The terms "revolution" and "paradigm shift" are reserved for those changes that do *not* measurably improve our understanding of the natural world.

The second biggest example of a paradigm shift is Einstein's 1905 special relativity theory. It has even been called the second scientific revolution. It was a revolution because it generated a lot of excitement, but it did not explain electrodynamics any better than the Lorentz-Poincare theory. There was not even any formula or prediction that was any different (except for mistakes and approximations). It was only different because Einstein explained the concepts in a slightly different order, and used slightly different terminology. Einstein's explanation was also more accessible to those physicists with limited mathematical knowledge. Kuhn considered it revolutionary because Einstein did such a poor job of relating it to previous work.

While these works of Copernicus and Einstein are universally praised as the greatest in the history of science, they actually had no significant influence on scientific progress. Heliocentric astronomy developed from Aristarchus, Ptolemy, Tycho, and Kepler, without any help from Copernicus. Special relativity developed from Maxwell, Lorentz, Poincare, and Minkowski, without any help from Einstein.

Kuhn hated Popper's falsification concept because these paradigm shifts did not falsify the previous theories. Copernicus did not falsify Ptolemy, and Einstein did not falsify Lorentz. Furthermore, the observational errors of the Copernican model could be fudged by adjusting or adding epicycles. Falsification is useful when there is some observation that shows that one theory is demonstrably superior to another, but is not useful for distinguishing between two theories that predict the same observations.

Thus Kuhn's notion of a paradigm shift does not mean a superior theory; it means an irrational change from one theory to another. (Kuhn preferred the made-up term *arational*, to mean not rational, without necessarily being contrary to reason.) A revolution means a change in a point of view, without necessarily any substantive change in anything real. The new paradigm is *incommensurable* with the old one, so there is no way to objectively compare the new to the old and give some measurable advantage to the new one. The new paradigm wins out because the young scientists suddenly adopt it and the old ones die out. Kuhn's favorite example is the Copernican model, which he suggests that astronomers adopted out of Sun worship rather than scientific merit.[356]

Kuhn's greatest popularity is among the soft sciences, yet he was only concerned with the hard sciences. He looked at theories that made hard quantitative predictions, and his paradigm shifts were to newer theories that also made quantitative predictions. But to be a paradigm shift, those newer theories cannot have any objective quantitative advantage over the older theories. The new theory takes hold quickly because it is fashionable, not because it is better.

Soon after Kuhn's book was published, geology had what many people called its great paradigm shift in the 1960s. The theory of *continental drift* became adopted, and renamed *plate tectonics*. It revolutionized geology in the sense that a new idea[357] was quickly and universally accepted, and it transformed the field by explaining mountain ranges and many other geological features. But it was not like the Copernican revolution. Young and old scientists were persuaded by hard scientific evidence that the new theory was superior. It wasn't just a change in a point of view. It was not even a change in methodology. Either the continents were once connected or they were not, and the 1960s produced overwhelming evidence that they were. Likewise the discovery of the molecular structure of DNA and its role in genetics had a dramatic effect on biochemistry, and might be called a paradigm shift. But Kuhn had no interest in this change either, as it was nothing like the sort of revolution that he was describing.

Since the term "paradigm shift" is used and misused so broadly, it is better to use the term *Kuhnian paradigm shift* for one that is incommensurable and not rational, as Kuhn describes. Thus De Revolutionibus was a Kuhnian paradigm shift, but plate tectonics theory was not. Einstein's 1905 theories of photons and special relativity were Kuhnian paradigm shifts from the Planck and Lorentz-Poincare theories, because Einstein's work had no rational or measurable advantages over the previous work. Kuhn later wrote a whole book on the early development of quantum mechanics, without claiming that it was a paradigm shift.[358] To Kuhn, Planck's discovery of the quantization of light was not a paradigm shift, but Einstein's restatement of the same basic idea five years later might be.

Planck discovered a new fundamental constant h, now called Planck's constant. It soon became as important as the speed of light. As the speed of light defines the symmetry between space and time, Planck's constant defines the quantum mechanical symmetry between position and momentum. Planck's constant gives the energy and momentum of photons and the spin angular momentum of electrons. It defines the boundary between classical and quantum mechanics. It had immediate applications to explaining puzzling radiation experiments. It was a profound and important breakthrough in physics, any way you want to measure it. And yet, because the advantages of the Planck theory were so directly measurable, it was not a paradigm shift. A Kuhnian paradigm shift must be incommensurable with previous theories.

Kuhn got the idea of incommensurability from a Greek legend about Pythagoreans being upset about the discovery of the irrationality of the square root of 2. Supposedly the cult was upset that no simple integer or fraction could be squared to get 2, even though such a number seems to be needed for the length of a hypotenuse. The apocryphal story is that they murdered the discoverer out of fear that their philosophy of explaining the world with numbers would be exposed as fallacious.[359]

While it is true that the square root of 2 has an infinite non-repeating decimal expansion, this fact does not limit Pythagorean philosophy or any later science. In spite of the importance that Kuhn and other philosophers attach to incommensurability, there is no known example in science where the concept is actually useful. Numbers are vital in the hard sciences for measuring how one theory is better than another, and today's philosophers have trouble understanding what the Greek thought was obvious 2500 years ago. The British physicist William Thomson (Kelvin) is known for discovering absolute zero temperature, and he explained in 1883:

> When you can measure what you are speaking about, and express it in numbers, you know something about it; but when you cannot measure it, when you cannot express it in numbers, your knowledge is of a meager and unsatisfactory kind: it may be the beginning of knowledge, but you have scarcely, in your thoughts, advanced to the stage of Science, whatever the matter may be.[360]

Dyson denied that he was a Kuhnian, because Kuhn's theory did not match the progress of particle physics in his lifetime. But he declared that "Kuhn was right" about paradigm shift theory explaining the history of special relativity.[361] Dyson said that Poincare and Einstein had the same theory, except for obscure terminological differences, and they were "unequal only in their receptiveness to new ideas." But Dyson complained that Einstein was not getting enough credit in Galison's relativity history book because "Einstein was by temperament revolutionary" and blamed Poincare for trying to relate his theory to previous work by others. Dyson's argument was that Einstein led a Kuhnian paradigm shift precisely because his work had no measurable or rational advantages. Dyson's attitude is common among theoretical physicists. Few of them will admit to believing in Kuhn's more provocative ideas, but they will usually give Kuhnian arguments for crediting Einstein with special relativity. Their arguments for Einstein are not based on any objective reality, or on any measurable progress towards truth. They have become Kuhnians without realizing it.

The striking thing about Kuhnian paradigm shift theory is that it attaches such great importance to events that have no great objective significance. It describes the history of science in terms of very important-sounding revolutions, and then denies that any of these revolutions advanced us towards truth in any measurable way. It gives the impression that science just jumps from one fad to another without any rational basis or real progress.

The British philosopher Bertrand Russell said:

> No opinion should be held with fervour. No one holds with fervour that seven times eight is fifty-six, because it can be known that this is the case. Fervour is only necessary in commending an opinion which is doubtful or demonstrably false.

And so it is with paradigm shifts. No one argues with any fervor about the great scientific advances of Faraday, Pasteur, Maxwell, Lorentz, and Poincare. They argue with great fervor that Galileo proved the motion of the Earth or that Einstein created relativity theory or that there is no aether or that there is intelligent life on other planets.

Anyone with a cell phone can see that science has made progress. The progress in the last century is astounding and undeniable. This paradigm shift theory would hardly be worth mentioning except that it is overwhelmingly accepted among intellectuals and in universities, especially in the soft sciences and humanities.

Extremist followers of Kuhnian paradigm shift theory deny that there is any objective reality or that science can tell us any objective truths. More commonly, professors make milder statements such as this:

> Yet, on the other hand, science must proceed in a social context and must be done by human beings enmeshed in the constraints of their culture, the throes of surrounding politics, and the hopes and dreams of their social and psychological construction. We scientists tend to be minimally aware of these human influences because the mythology of our profession proclaims that changing views are driven by universal reasoning applied to an accumulating arsenal of observations. But all scientific change is a complex and inseparable mixture of increasing knowledge and altered social circumstances.[362]

No, objective science is not a myth. The myths are the distortions in the history of science that have allowed paradigm shift theory to flourish. Copernican astronomy was no great advance, and neither was Einstein's special relativity paper. They were only revolutions in the sense of offering a slightly different worldview on previous work. Much more important astronomy work was done by Ptolemy, Tycho, and Kepler, and much more important electrodynamics work was done by Maxwell, Lorentz, and Poincare. These real advances were objectively and quantitatively measured by experiments, and accepted by the experts of the day. The history of science is a history of objective progress, in the hard sciences at least.

Quantum leaps

Another confusing and misused science cliché is the *quantum leap*. Like the paradigm shift, it is often used to suggest some sort of sudden and radical change from the past. The German car-maker BMW has an advertising campaign that says: "A quantum leap is defined as a dramatically large advance, especially in knowledge or method. … experience this quantum leap".[363]

The phrase comes from early 20th century physics. A quantum leap is a tiny *discrete* jump from one energy level to another. The theory of quantum mechanics teaches that many physical quantities are only observable in discrete amounts. The *quantum* is the smallest non-zero magnitude possible. For example, the electron has a (negative) quantum of electric charge, as all other charges are integer multiples of an electron charge. The electron also has a quantum of spin angular momentum, one half *h-bar*. Light energy from a laser beam is always an integer multiple of hf, where h is Planck's constant and f is the frequency of the light. That is why we say that light is quantized into photons. It is like saying that the quantum of American money is the penny. All American money is a multiple of the penny.

An electron in an atom will only exist at certain discrete (i.e., discontinuous) energies, and it will sometimes make a discontinuous jump to another energy. It absorbs or emits a photon of a particular frequency in the process. The frequency is also the color, if the light is visible. Modern *spectroscopy* is based on measuring the frequencies of those photons. Because the frequencies are discrete and different for different atoms, spectroscopy can tell you what atoms are in a lab sample. The method is amazingly precise and useful. Quantum mechanics gives formulas for those electron energies, as well as probabilities for an electron to jump to a nearby energy level. This is the core of what made the theory so successful in the 1920s. But when an electron makes a quantum leap, it is not dramatically large, and it is not an advance. Typically, the electron is just randomly bouncing back and forth between two adjacent energy levels. The leap is as tiny as it could possibly be.

It is not known that nature itself is so discrete. It is possible to interpret quantum mechanics in terms of waves and fields, instead of particles. Schroedinger wrote in 1952 that there is no such thing as a quantum jump in nature.[364] It just appears that way because observations cause sudden changes to our knowledge about a system. Most of the other leading physicists preferred interpretations in terms of particles,[365] but there is no way to prove that any one interpretation is more correct that any other.

The quantum leap is not just an overused metaphor, it is a way to put a pseudoscientific veneer on subversive ideas. The German philosopher Georg Wilhelm Friedrich Hegel wrote in the early 1800s about a qualitative leap (ein qualitativer Sprung) in consciousness. The German communist gurus Karl Marx and Friedrich Engels adopted Hegel's terminology about dialectical materialism. They pretended to have a scientific theory about how historical events are the dramatic and inevitable consequences of dialectical law. They used this theory to justify revolutions and sound more scientific. Here is a typical Marxist discussion of a quantum leap:

> The dialectical method seeks to explain natural phenomena as the transformation of quantity into quality: a long period of slow, gradual change is interrupted at a critical point by a sudden change of state, a quantum leap, a phase transition or, to use the language of dialectics, a qualitative leap. This method of analysis was first developed by Hegel two hundred years ago and then placed on a scientific basis by Marx and Engels. But it is only in recent years, thanks to the development of chaos theory and its derivatives that it has begun to be taken seriously by scientists.[366]

Real scientists do not take this nonsense seriously. The scientific basis for the quantum leap does not support the Marxist agenda. It makes more physical sense for the Marxists to make analogies to phase transitions, such as when water boils or freezes. Then a substance suddenly changes between the solid, liquid, and gas states. But then their theory is less impressive, because phase transitions have been known for thousands of years, and they do not convince anyone that we are in need of a Marxist revolution.

While some people are always looking for quantum leaps, scientists since Aristotle have followed the slogan, *Natura non facit saltus* (Latin for "nature does not make jumps"). Great discoveries by Newton, Darwin, and many others were based on a conviction that substantial changes can be described by a gradual accumulation of tiny changes. Maxwell and Poincare said that the aether was invented so that the light from the stars could be explained as a series of propagating waves.

Some say that quantum mechanics changed all that, but here is what Heisenberg wrote in 1958:

> When old adage 'Natura non facit saltus' is used as a basis of a criticism of quantum theory, we can reply that certainly our knowledge can change suddenly, and that this fact justifies the use of the term 'quantum jump'.[367]

He was not saying that nature makes the jump, but rather that our knowledge of nature changes suddenly when we make an observation. We could open a box and suddenly discover that Schroedinger's cat is dead. It may have been uncertain whether the cat was already dead. To Heisenberg, the fancy mathematics of quantum mechanics was really just describing our knowledge of particles and fields, and those quantum leaps were just sudden changes in that knowledge.

Many areas of science have those who emphasize continuity theories, and those who look for theories of sudden change. Ideas of sudden change in biology go under fanciful names like *hopeful monster* and *punctuated equilibrium*. The historians of science are divided between those who trace ideas back to ancient times, and those who think that Copernicus led a sudden scientific awakening.

Continuity theories have always been much more successful. If a continent is drifting at a rate of two inches per year, and has been doing so for the last 100 million years, then it is an easy extrapolation to predict that the continent will continue to drift similarly for the next million years. But some of that movement may happen with a sudden earthquake on a particular day, and that quake is nearly impossible to predict. An earthquake is an example of a sudden movement in nature, but it is still best understood in terms of being caused by continuous forces.

It would great if some future scientific theory lets us predict those quantum leaps, earthquakes, stock market crashes, political revolutions, and other sudden changes. But scientific predictability is almost always based on continuity and causality. And that includes quantum mechanics, which uses continuous functions and differential equations to describe all those leaping electrons.

Poincare's conventionalism

Poincare had his own philosophy of science called *conventionalism.* Popper considered Poincare to be the greatest philosopher of science ever.[368] In his 1902 book, *Science and Hypothesis,* Poincare said:

> These are the questions which naturally arise, and the difficulty of solution is largely due to the fact that treatises on mechanics do not clearly distinguish between what is experiment, what is mathematical reasoning, what is convention, and what is hypothesis.

These misunderstandings are still common in textbooks today. Michelson-Morley and eclipse starlight deflection were experiments. The Pythagorean theorem in Euclidean geometry is proved with mathematical reasoning. A hypothesis was that the aether velocity was unobservable. Using the metric system of measurement is a convention, and so is choosing a system of units with the speed of light equal to one. Where Poincare's view was unusual was that he argued that various scientific facts were really conventions. It is just convention that we say that the Earth goes around the Sun because it is possible to have a theory of the Sun going around the Earth and it would be just as consistent with our experience. Einstein once said that he agreed with Poincare's conventionalism.[369]

In a popular 1908 book, Poincare applied his distinctions to relativity:

> This hypothesis, formulated by Lorentz and FitzGerald, will at first seem extraordinary; all we can say in its favor at the moment is that it is only the immediate translation of the experimental result obtained by Michelson, if we define lengths by the time light takes to traverse them.

In this sentence, the experiment is Michelson-Morley, the convention is the definition of length in terms of light, the hypothesis is the length contraction, and the reasoning is the logic that leads to that hypothesis from those premises. And yet a recent article in a historical physics journal[370] quoted this very paragraph and cited it as proof "that Poincaré's thinking stopped short of the crucial step" in understanding special relativity. The article objects to part of relativity being only a "hypothesis", and the author does not accept the distinctions that Poincare was making. Mathematicians tend to be much more precise about such distinctions than physicists.

To a mathematician, Poincare's distinctions are like the distinction between facts and opinion. It is difficult to have an intelligent debate with someone who does not accept that distinction. If someone tells you that you are wrong about something, you first want to determine whether he is claiming that your facts are wrong, or if he just disagrees with your opinions. Likewise, if you were disputing Poincare, he would want to know whether you were challenging his experiments, reasoning, conventions, or hypotheses. If you confuse these concepts, then you are going to have a hard time understanding Poincare.

The American physicist Steven Weinberg once wrote a book on general relativity[371] in which he de-emphasized the curvature of spacetime. He preferred to think of a flat spacetime with some gravitational fields that happen to have an interpretation as curvature. His preface said that, "I believe that the geometrical approach has driven a wedge between general relativity and the theory of elementary particles", and that "Riemannian geometry appears only as a mathematical tool". Other astrophysicists treated him as if he had missed the whole point of general relativity. But he was just subscribing to Poincare's conventionalism, and it is impossible to say that he was wrong. His text used the same analytic and geometric methods to get the same results as other relativity textbooks. He was just adopting a view that others consider inconvenient.

Weinberg's preface was like the infamous Osiander forward to Copernicus's book. Both were making the point that a mathematical theory about the physical world can have more than one interpretation. There will always be realists who insist that just one interpretation is the correct description of reality, but even with that view, those other interpretations may also be useful to those who wish to subscribe to them. There are many examples in modern physics where we are unable to specify the underlying reality to everyone's satisfaction.

Of course some conventions really are superior to others, and there may be a public consensus that a particular convention reflects reality. For example, most physicists would say that the Earth really does go around the Sun, and spacetime really is curved. Or that the Earth goes in a straight line in spacetime. Others are more empiricist, and say we can only be sure about the outcomes of experiments.

When Poincare said that the aether was a "convenient hypothesis", he wasn't saying anything about whether the aether physically exists or whether he believes in it. He merely meant that speaking about the aether was a useful convention for discussing the propagation of electromagnetic waves. Whether or not the aether was observable in experiments is an entirely separate question. He was sometimes accused of having a theory that depended on an aether, or a privileged frame, or true time,[372] but that is a gross misunderstanding of his conventionalism. He proved that his theory was independent of these concepts by constructing a symmetry group, and he did not even mention them when he described his theory.

There are many examples of mathematically equivalent theories being used to describe physical phenomena, and in those examples, choosing one is a matter of convention. An ancient example is the equivalence between geocentric and heliocentric models of the solar system. A modern one is quantum field theory, which can be formulated in terms of either particles or fields. Another is the interpretation of quantum mechanics, which can be in terms of observers or multiple universes. People can get very excited about which should be regarded as more correct, but unless some experiment can show that one interpretation is better than the other, the choice is ultimately based on whatever is more convenient.

It was once thought that there were two contradictory hypotheses about the nature of light. It could be composed of waves or particles, and experiments would determine which is correct. But that is now known to be an oversimplification. The wave and particle theories of light are both convenient hypotheses, and both theories are extremely useful for predicting experiments. Likewise, an electron or a quark can be considered a particle or a wave. A realist might argue that an electron really is a particle or a wave, and that we should adopt a theory that we believe to conform to reality. Bohr said that the opposite of a correct statement is a false statement, but the opposite of a profound truth may well be another profound truth.[373] For Poincare, the prediction of experiments is the ultimate test of a scientific theory, and a theory may depend on conventions that are not directly testable.

Poincare said that a hypothesis should always be tested, in order to try to verify it. It must be immediately abandoned if it fails the test. He said this as if it were conventional wisdom, and it was similar to what Popper later said about falsification. Some hypotheses are difficult to test, for various reasons. For example, the Greek hypothesis that matter consists of atoms went untested for two millennia. Other hypotheses, such as the existence of the aether, might be convenient even if they cannot be verified.

Sometimes equivalent theories are accepted as equally valid. Newtonian, Lagrangian, and Hamiltonian mechanics are all equivalent, and no one tries to argue that one is more correct than another. Heisenberg and Schroedinger developed equivalent formulations of quantum mechanics. Quantum electrodynamics has been formulated in terms of particles that act like fields and in terms of fields that look like particles, and physicists agree that they are equivalent because they make the same predictions.

Modern physics theories often have multiple interpretations. For example, the standard model teaches that quarks are fundamental particles but they can never be isolated. So are the quarks real particles, or just mathematical constructs that help us make predictions about nuclear physics? Most physicists would say that the quarks are real, because the theory is just too good. But modern physics is filled with little conundrums like this. If you ask whether alternate universes exist, you will get all sorts of answers. One famous physicist once said, "I do take 100 percent seriously the idea that the world is a figment of the imagination."[374]

Others say that the universe could be one gigantic computer simulation, as in the 1999 movie *The Matrix*. Or that we are the invention of a Boltzmann brain, which is a hypothetical self-aware entity arising from some natural energy fluctuation. Or that any mathematically consistent set of laws defines some universe, and we just happen to live in one of the more interesting ones. Ancient Greek philosophers such as Plato and the solipsists made arguments similar to these. Such ideas are hard to refute because they have no observable consequences.

The point here is that if there is no experiment or observation to distinguish competing theories, then there is no way to say for sure that one theory is any more correct than another. In such a situation, Poincare would say that the physicists should leave such questions to the metaphysicians.[375]

Hawking's 2010 book says that the Copernican system is no more real or correct than the Ptolemaic system. It explains and endorses conventionalism this way:

> When such a model is successful at explaining events, we tend to attribute to it, and to the elements and concepts that constitute it, the quality of reality or absolute truth. But there may be different ways in which one could model the same physical situation, with each employing different fundamental elements and concepts. If two such physical theories or models accurately predict the same events, one cannot be said to be more real than the other; rather, we are free to use whichever model is most convenient.[376]

Conspicuously absent from Poincare's philosophy is *Occam's Razor*. That is the idea that the simplest explanation should be regarded as the correct one. Poincare was a brilliant mathematician, and sometimes one explanation seems simpler because it uses less sophisticated mathematics. But it might not be any simpler to Poincare, so he might regard it as an equally attractive alternative. When he reformulated electromagnetism in terms of a relativistic action, no one thought that it was any simpler, but the formulation ultimately became extremely important for quantum field theory decades later, and it is simpler for some purposes.

The Copernican heliocentric theory was more or less observationally indistinguishable from its geocentric rivals in the 1500s. Some say that the Copernican system was simpler, but it really was not. They all could be considered just different interpretations of the same underlying theories. One of them might be conceptually more satisfying, or might allow easier calculations, or might be easier to explain in a textbook. Choosing one was a matter of convenience.

Kepler wrote that astronomy has two purposes. The first is to describe the appearance of the sky and the second is to explain the true form of the universe.[377] Having a theory that is consistent with observation was sometimes referred to, by ancient philosophers, as *saving the ap-*

pearances, or *saving the phenomena*. Nowadays it is sometimes called *phenomenology*. Either a geocentric or heliocentric theory could do that, and satisfy Kepler's first purpose for astronomy. Kepler was very concerned with the second purpose also, but his arguments were partly philosophical. He stopped short of claiming that he could prove the motion of the Earth. Ptolemy believed in this same dichotomy in the purposes of astronomy. He separated them into two books, with his Almagest describing the appearance of the sky, and his Planetary Hypotheses trying to explain the true form of the universe.

While Poincare invented the 4-dimensional spacetime view in 1905, he did not argue that it was a necessity as Minkowski did in 1908 and as Einstein did a few years later. Here is what Poincare said about it in a 1912 lecture shortly before his death:

> The new conception ... according to which space and time are no longer two separate entities, but two parts of the same whole, which are so intimately bound together that they cannot be easily separated... is a new convention [that some physicists have adopted]... Not that they are constrained to do so; they feel that this new convention is more comfortable, that's all; and those who do not share their opinion may legitimately retain the old one, to avoid disturbing their ancient habits. Between ourselves, let me say that I feel they will continue to do so for a long time still.

His point was that you can think of space and time as being separate or as being unified. Either way, you would have to accept the spacetime symmetries, so there may be no practical difference. A century later, space and time are still distinguished for most purposes, as Poincare predicted. For example, your wristwatch tells time without making any reference to spacetime.

Relativity was a great example of Poincare's conventionalism. We think of space as being isotropic, meaning that all directions are alike. From the view of a moving observer, distances are contracted in the direction of the motion.[378] Because the Lorentz transformation is a spacetime symmetry, the isotropic and contracted views are equally valid. This raises the possibility that space is not really isotropic, but it just looks isotropic because of properties of our measuring devices. Space is taken to be isotropic as a matter of convenience.

Poincare argued that space could be deformed and we would never notice if our laws of physics took the deformations into account. The Lorentz transformation was proof of that. He even said that some definitions of energy are useful conventions that were concocted to perpetuate our belief in conservation of energy. For example, when you light a match, the flame appears to have energy that did not exist before. So chemical potential energy was defined to describe the energy that was contained in combustibles.

There were two major interpretations of Lorentz transformations in the special relativity of 1905. Einstein's interpretation was that they affect measuring rods and clocks, and that there is a way to extend them to electromagnetic variables so that Maxwell's equations take the same form in different reference frames. Lorentz had published the same interpretation in 1895, after FitzGerald conjectured a simplified version of it in 1889. Poincare's interpretation was that the Lorentz transformations are symmetries of spacetime, and that the relativity principle is a consequence of covariance under those symmetries. Within several years, Poincare's interpretation was adopted by Minkowski, leading physicists, Einstein, and relativity textbooks. Lorentz and Poincare themselves recognized that both interpretations are viable.

The existence of the aether was a matter of convention for Poincare, as no experiment could show any physical property of the aether at the time. You could assume that there was an aether, or that there was not an aether, depending on what was more convenient. He defended Fresnel's theory of optics, even though it was based on dubious aether assumptions.[379] The situation is analogous to a medieval astronomer switching between the Ptolemaic and Copernican models to predict the location of Mars in the sky. Both methods were about equally useful. Poincare is sometimes criticized for his ambivalence about such issues, as if he had failed to understand the conceptual superiority of one theory. But Poincare's ambivalence was rooted in an attempt to separate science from philosophy.

Historian Arthur I. Miller says that the main reason Poincare should not share the credit for special relativity with Einstein is that Poincare did not "elevate the principle of relativity to a convention". What Miller meant by this is that Poincare based the principle on experimental evidence, such as Michelson-Morley. If new evidence disproved the principle, then he would have to reject it.[380]

A convention is not some higher form of truth. It is the lowest form of truth. A convention cannot be proved by either mathematics or experiment. Someone who prefers not to adopt the convention cannot be proven wrong. Poincare's understanding of relativity was at a higher level than Einstein's, and it is to Poincare's credit that he correctly distinguished what was a matter of convention.

6. Physics becomes science fiction

Modern physics has been taken over by academic researchers who call themselves theoretical physicists but who are really doing science fiction. They are not mathematicians who prove their results with logic, and they are not scientists who test their hypotheses with experiments. They make grand claims about how their fancy formulas are going to explain how the world works, and yet they give no way of determining whether there is any validity to their ideas.

Psi physics gets spooky

The 20th century was the golden age of science. More fundamental science breakthroughs were made in that century than any other. Science will continue to make progress, but the basic discoveries have been made.

And yet there is a popular myth that the 20th century just created a lot of uncertainty and confusion in our scientific knowledge. Relativity taught us that we cannot be sure about motion or time; quantum mechanics taught us that the universe is probabilistic. Either way, the lesson seems to be that reality is unknowable, and that measurements just create more uncertainty. This impression is a mistake.

Here is an example of an attempt to draw philosophical consequences from modern physics. Harvard Law professor Laurence H. Tribe wrote a 1989 paper[381] that thanked his student (and future President of the United States) Barack Obama, and begin with this abstract:

> Twentieth-century physics revolutionized our understanding of the physical world. Relativity theory replaced a view of the universe as made up of isolated objects acting upon one another at a distance with a model in which space itself was curved and changed by the presence and movement of objects. Quantum

> physics undermined the confidence of scientists in their ability to observe and understand a phenomenon without fundamentally altering it in the process. Professor Tribe uses these paradigm shifts in physics to illustrate the need for a revised constitutional jurisprudence. ...

Tribe wrote:

> A second criterion for choosing among competing paradigms might be called the "progressivity" of the paradigm – the resilience and usefulness of the paradigm in a new context. A progressive paradigm adapts in a constructive fashion to new "data" – new situations and problems; a "degenerative" paradigm must be revised in an ad hoc fashion to handle these new facts or contexts. ... The Einsteinian paradigm is, in this way, more progressive than the Newtonian paradigm.

It took centuries of observations of the planet Mercury to even detect a slight difference between the Einsteinian and Newtonian paradigm, and yet this law professor is able to declare that the shift has leftist political implications.

Such applications of scientific ideas to politics go back centuries. After being elected the American President in 1912, Woodrow Wilson published a book of campaign speeches[382] including a *Copernican federalism* argument that "the Constitution of the United States had been made under the dominion of the Newtonian Theory." The American founding fathers, he said, "constructed a government as they would have constructed an orrery [planetarium] – to display the laws of nature." He argued that the goal of progressives was to interpret the American Constitution according to the Darwinian principle of adaptation, and not the Newtonian principle of blind forces.

The New York University physicist Alan D. Sokal has made a second career for himself by making fun of academic non-scientists who use physics metaphors inaccurately. He published a paper in 1996 that strung together a bunch of such quotes, and filled in jargon-heavy explanations. After the paper appeared in a Marxist journal, he claimed that it was some sort of hoax or parody, and that he did not really believe what he said in the paper. He explained that his purpose was to embarrass his fellow radical leftists into using better science to support their leftist political causes. He has since written books and articles explaining the matter further.[383]

Confused physics metaphors are not just used by law professors and deconstructionist philosophers. Sokal acknowledged that his fellow physicists use them with quantum mechanics:

> Let us also stress that in our book we have rigorously refrained from criticizing postmodernists for abuses related to quantum mechanics – otherwise, the book would be considerably longer – precisely because we feel that it would be unfair to criticize non-physicists on a subject where the physicists themselves are sometimes quite confused.[384]

There are indeed many goofy statements about quantum mechanics by reputable physicists, from the 1930s to today. Some will say that quantum mechanics requires consciousness, or that it denies objective reality, or that it is spooky, or that it requires action-at-a-distance, or that it makes anything possible. It ought to be unnecessary, but Weinberg has had to write essays explaining that he believes in objective reality.[385]

Even if the journal editors had sent Sokal's paper to an expert physicist for review, it is not clear that the outcome would have been any different. Denying objective reality does not raise eyebrows anymore. When Weinberg scrutinized the paper for the purpose of attacking misuses of science, he found surprisingly little wrong with what Sokal actually wrote. Weinberg's review said:

> I thought at first that Sokal's article in Social Text was intended to be an imitation of academic babble, which any editor should have recognized as such. But in reading the article I found that this is not the case. ... Where the article does degenerate into babble, it is not in what Sokal himself has written, but in the writings of the genuine postmodern cultural critics quoted by Sokal.[386]

Weinberg does complain about some "howlers", such as quoting, "the Einsteinian constant is not a constant", and being a little sloppy about how non-commutativity implies nonlinearity. (It is true that non-commutativity in a gauge group implies that the quantum fields are nonlinear, but Weinberg objects to the way that it is phrased.) But it is hard to see how these physics metaphors would confuse anyone. Physicists often say similar things. Dirac once said, "perhaps the gravitational constant is not really constant at all."[387] You would much more likely be confused by the non-science postmodernist mumbo-jumbo.

Sokal claimed that he "intentionally wrote the article so that any ...undergraduate physics or math major would realize that it is a spoof." But he seemed to be perpetrating another hoax by saying that, as today's undergraduates are subjected to goofier ideas from respectable scholars. His paper did not say that the Moon was made of green cheese, or anything so obviously ridiculous.

There are many non-physicists who invoke the mysteries of quantum mechanics in order to promote mystical ideas. For example, the New Age guru Deepak Chopra has written dozens of books on using the mind for alternative medicine. He cites quantum mechanics to support the idea that human consciousness can eliminate aging and illness.[388] More goofy quantum mechanical ideas were presented in the popular documentary-style 2004 movie, *What the Bleep Do We Know!?* It tried to relate physics to an assortment of spiritual beliefs.

If you want to read nonsensical applications of physics, it is easier to just read what the physicists are saying themselves about quantum mechanics. They talk about spooky action-at-a-distance, half-dead cats, quantum teleportation, and alternate realities. While there is spectacular experimental confirmation for the wave properties of particles and for the standard model, there is no confirmation at all for these ideas.

In a televised 2010 debate on the scientific future of God,[389] Chopra addressed the reality of the Moon by saying, "In the absence of a conscious entity, the Moon remains a radically ambiguous and ceaselessly flowing quantum soup." Caltech physicist (and part-time science fiction writer) Leonard Mlodinow said that he was writing a book with Hawking, and challenged Chopra to learn some quantum mechanics. But that book talks about non-locality and God just like Chopra, and it says this about the Moon:

> There might be one history in which the moon is made of Roquefort cheese. But we have observed that the moon is not made of cheese, which is bad news for mice. Hence histories in which the moon is made of cheese do not contribute to the present state of our universe, though they might contribute to others. That might sound like science fiction, but it isn't.[390]

Apparently Chopra subscribes to the Copenhagen interpretation of quantum mechanics, while Hawking and Mlodinow subscribe to the alternative histories interpretation. Cosmologists do not like the Copenhagen interpretation because it seems to depend on human consciousness, and that makes it harder to speculate about what happens in other galaxies. But there is no proof that any one interpretation is any more correct than any other, and they are all jumping to untestable conclusions.

The big physics breakthroughs of the 20th century were the discoveries of new symmetries of nature. Relativity found symmetries between space and time, between electricity and magnetism, between momentum and energy, and among frames of reference. Quantum mechanics found symmetries between position and momentum, between time and energy, and in a spin group. Gauge theories found symmetries about a circle bundle for electromagnetism, and other bundles for weak and strong interactions. Particle physics found various other broken symmetries.

Finding a symmetry of nature does not cause uncertainty. The symmetry reduces the uncertainty. The Heisenberg uncertainty is only an uncertainty if you think of electrons as Newtonian particles that bounce around like billiard balls. But they are not. Electrons are really wave-like objects that sometimes look like particles. They cannot be localized like simple particles because they are not simple particles. A musical note cannot be so localized for the same reason.

When Michelson and others discovered that light behaves the same in different moving frames, we learned more about nature, not less.

Uncertainty did not start with quantum mechanics. Poincare discovered that celestial mechanics can become chaotic, and that was before quantum mechanics was discovered. Chaos theory does not require quantum mechanics. The Earth's weather would be chaotic under classical or quantum mechanics. The universe is not deterministic in any practical sense whether you use 19th century or 20th century physics.

The theory of quantum mechanics seems more unpredictable because it has unobservable variables. A basic object is the wave function, which is traditionally denoted with the Greek letter psi. It is interpreted in terms of probability, which is not directly observable. Position and momentum are not simultaneously measurable precisely, because of Heisenberg uncertainty, so they are not jointly observable. But earlier theories have had unobservable variables also. Lagrangian and Hamiltonian mechanics can use mathematical variables that do not have direct physical interpretations. So did Maxwell's electromagnetism.

The theory of black holes gives the most striking example of an unobservable. We can see matter falling into the event horizon and we can see the gravitational effects of a black hole, but the inside of a black hole is fundamentally unobservable. Light cannot get outside the event horizon. Relativistic causality teaches that if light cannot get out, then nothing inside a black hole can have any causal effect on anything outside. It is not clear that it even makes any scientific sense to talk about what is inside the event horizon, because there can be no test to show whether any such ideas are correct or not. The mathematical physics predicts a singularity at the center of a black hole, where the diameter and density are infinite. Such infinities are hard to understand physically, but they are not observable anyway because they are inside the event horizon.

As successful as quantum mechanics is, no one knows what it all means. Feynman probably understood it better than anyone, and he said that he didn't understand it. The formulas correctly predict experiments to as many as eight decimal places, but it is very hard to get a conceptual understanding of what is really going on. It is just too weird. Even Dirac scoffed at the successes of quantum field theory, and said, "Just because the results happen to be in agreement with observation does not prove that one's theory is correct."[391] We cannot come to firm conclusions about what quantum mechanics means because there are multiple interpretations of it. There is no consensus on which interpretation is correct, and they all agree with experiment, as far as we know.

Nevertheless it is very misleading to say that the world is a strange and unpredictable place because of quantum mechanics. It would be stranger and less predictable without quantum mechanics. The theory explains why all carbon atoms have the same chemical properties. That fact is essential to all life on Earth, and there is no way to understand it without quantum mechanics.

Nearly all popular explanations of the strangeness of quantum mechanics are based on some particular interpretation, without even mentioning the fact that there are other equally valid interpretations.

So all they are really saying is that one particular interpretation of the equations has some strange properties. The result is very misleading because the interpretations are stranger than the theory itself, and a strange feature of one interpretation is not necessarily shared by other interpretations.

A recent physics paper predicts that time will end in the next five billion years.[392] The argument is that eternal inflation predicts infinitely many universes, with anything being possible in those universe, making probabilities confusing unless it all comes to an end in finite time. The argument is silly, but no one suspects a hoax. Arguments like this have become commonplace in theoretical physics.

A recent conference on alternative interpretations of quantum mechanics attracted some physicists with an interest in the paranormal, including one physics Nobel prizewinner who once wrote a paper with this abstract:

> A model consistent with string theory is proposed for so-called paranormal phenomena such as extra-sensory perception (ESP). Our mathematical skills are assumed to derive from a special 'mental vacuum state', whose origin is explained on the basis of anthropic and biological arguments, taking into account the need for the informational processes associated with such a state to be of a life-supporting character. ESP is then explained in terms of shared 'thought bubbles' generated by the participants out of the mental vacuum state. The paper concludes with a critique of arguments sometimes made claiming to 'rule out' the possible existence of paranormal phenomena.[393]

Physicists tolerate some very strange views on the subject of quantum mechanical interpretations, but the paranormal is on the fringe. There was some controversy about whether such views should be allowed to be expressed at the conference. Since then, a respectable psychology journal has published a paper purporting to show evidence for a form of ESP called precognition, and citing dubious quantum mechanical reasoning:

> The psychological level of theorizing just discussed does not, of course, address the conundrum that makes psi phenomena anomalous in the first place: their presumed incompatibility with our current conceptual model of physical reality. Those who follow contemporary developments in modern physics,

> however, will be aware that several features of quantum phenomena are themselves incompatible with our everyday conception of physical reality. Many psi researchers see sufficiently compelling parallels between these phenomena and characteristics of psi to warrant considering them as potential candidates for theories of psi. ...
>
> Unfortunately, even if quantum-based theories eventually mature from metaphor to genuine models of psi, they are still unlikely to provide intuitively satisfying mechanisms for psi because quantum theory fails to provide intuitively satisfying mechanisms for physical reality itself. Physicists have learned to live with that conundrum but most non-physicists are simply unaware of it; they presume that they don't understand quantum physics only because they lack the necessary technical and mathematical expertise. They need to be reassured.[394]

The paper then goes on to quote Feynman on how reality can be hard to understand. Many scientists are dismayed by publication of papers like this, while others acknowledge that a lot of papers are wrong anyway.

String theory is not even wrong

The standard model of particle physics may be the greatest accomplishment of 20th century physics, but it gets no respect. The American physicist and science popularizer Michio Kaku calls it the "ugliest theory known to science." Stanford physicist Leonard Susskind calls it an "ugly monstrous mess."[395] Most people have never even heard of it.

The chief complaint about the standard model is that there are about 20 underdetermined real parameters in it. Most of these are particle masses for the quarks and leptons, which have to be measured experimentally. A theory that truly met Einstein's vision of a unified field theory, some physicists argue, would not have any such parameters. A truly unified field theory, they say, would derive all of the laws of the universe from postulates, and not depend on any arbitrary parameters. In particular, the particle masses would be deducible from pure theory.

For the past 40 years, theoretical physicists have been proposing unified field theories to replace the standard model. The merits of these unified field theories are always argued on aesthetic grounds. They have never agreed with experiment as well as the standard model, and they always have many more undetermined parameters. The theories cannot even be fully specified because of those unknown parameters.

So what is the appeal of these so-called unified theories? They are no simpler than the standard model, and they have no better predictive power. They all predict dozens of bizarre particles that have never been observed. (The standard model predicts only one, the Higgs boson.) Weinberg explained the aesthetic advantage: "all attempts to go beyond the standard model play with symmetries, some of which are broken and some of which are not."[396]

The belief that there should be a unified theory of everything has a history that goes back millennia. Ancient Greek philosophers before Plato often believed in *monism,* although they did not agree on the unifying concept. For some it was water, air, fire, or God. Eastern philosophies have similar ideas, and so do many religions. The desire for a unified field theory is a modern manifestation of this ancient belief.

Sometimes Einstein is credited with having created a unified field theory. A leading dictionary describes him this way:

> Einstein is generally regarded as one of the very greatest scientists in history. His ideas and speculations have brought about the most profound revolution in scientific thought since Copernicus. In 1905 he published four great discoveries in theoretical physics: the special theory of relativity, ... In 1950 he introduced a merger of quantum theory with the general theory of relativity, thereby establishing one set of determinate laws for subatomic phenomena and large-scale physical phenomena.[397]

No, Einstein did not discover the special theory of relativity, he did not merge quantum theory with the general theory of relativity, and he did not establish determinate laws for subatomic phenomena. His attempts at a unified field theory were misguided and fruitless. Nevertheless, he serves as the main inspiration for the search for a unified field theory. The researchers are always talking about how they are trying to complete the Einsteinian revolution.

The concept of the symmetry group has been such a powerful and fruitful idea for 20th century physics that physicists look for them everywhere. Of course, symmetries cannot be everywhere or else everything would look like everything else. So when they cannot find true symmetries, they look for broken symmetries.

According to the big bang theory, the entire universe was once the size of a peanut, and exploded rapidly. It is believed that the universe was pure energy at that time, and every form of energy was just like every other form of energy. Quarks, electrons, and photons were all the same. It was symmetry heaven for those who like symmetries.

The grand unified field theories all predict that the proton is unstable, and will eventually decay into lighter particles. If a proton is symmetrical with other particles at high energies, then there is some chance that it can be converted at low energies as well. Experiments have failed to find any such decay. None of the grand theories' other predictions have been verified either.

The grand unified field theorists are undeterred, and insist that they are still on the path to realizing Einstein's vision. The most downloaded physics paper is a grand unified field theory.[398] It is supposed to explain all the particles and forces, but it does not even have a left-handed fermion, so it cannot possibly explain the neutrino or any parity violating experiment. Now the theorists have moved on to string theory, which makes no predictions at all.

String theory is a proposal to unify all of the fundamental forces and particles into a single unified theory. It was considered so promising that it has dominated theoretical physics research over the last 25 years. Unfortunately it has been a colossal failure.

String theory replaces particles in spacetime with tiny strings in a spacetime with 6 or 7 extra dimensions. Supposedly it includes gravity and quantizes general relativity into a quantum theory. But none of its goals have ever been realized. Nobody has ever figured out any relation to electrons, photons, or any other known particles. No one has even formulated any equations of motion. No one has figured out how to relate it to gravity or any other known force. The closest that it comes is that string theory has a spin 2 massless particle, and since quantized gravity is conjectured to have a *graviton* of spin 2, some people are hoping that these will be the same. But string theory masses do not cause spacetime curvature, as is essential to general relativity.[399] No one has shown that it can even result in a consistent theory. Even some of the string theory leaders are now acknowledging that string theory may never have any predictive power.

Kaku describes string theory this way:

> Einstein said that the harmony he sees could not have been an accident. ... I work in something called String Theory which makes the statement that we are reading the mind of God... The Universe would be a symphony of these vibrating strings and the mind of God that Einstein wrote about at length would be cosmic music resonating through this nirvana... through this 11 dimensional hyperspace — that would be the mind of God. We physicists are the only scientists who can say the word "God" and not blush.[400]

Kaku describes a lot of fanciful ideas without blushing. God is Kaku's convenient hypothesis. In an interview[401] he explained, "We want a one-inch equation that would explain everything from the big bang to the creation of life and the universe as we know it." When asked whether we already have theories for those things, he responded that "God has two hands", and that Einstein's goal was to find one fabric that would unite them into one cosmic framework.

Part of the motivation for a unified field theory is a desire to eliminate the need for God. If some sort of final theory can erase the boundaries to human knowledge about the laws of physics, then there will be no room left for God. Each major new theory has led to a demystification of the heavens.[402]

Dutch string theorist Erik Verlinde proposed in 2010 that gravity is just an illusion caused by the dissipation of information. "We've known for a long time gravity doesn't exist," he said.[403] He was inspired by analogies to holographic images of exploding black holes. He defends his idea by saying that he is looking for the mathematics to prove his intuition, and that Einstein started out similarly.[404] His paper may turn out to be the most talked-about physics paper of 2010, if you can even call it a physics paper.

String theory is the most over-hyped theory of physics in a very long time. A recent congressional hearing declared:

> Unification was Einstein's great, unrealized dream, and recent advances in a branch of physics known as string theory give hope of achieving it. Most versions of string theory require at least seven extra dimensions of space beyond the three we are used to. The most advanced particle accelerators may find evi-

> dence for extra dimensions, requiring a completely new model for thinking about the structure of space and time…
>
> Understanding the very early formation of the universe will require a breakthrough in physics, which string theory may provide.[405]

No, the particle accelerators will not find evidence for extra dimensions. String theory does not make any predictions that there is any known way of testing. Dutch theoretical physicist Martin Veltman said, "String theory is mumbo jumbo. It has nothing to do with experiment."[406]

String theory is not even science, under Popper's falsification analysis. Even the string theory advocates deny that any experiment could ever prove string theory wrong. Caltech physicist Sean Carroll said, "the only way for someone to kill string theory will be to come up with a better one."[407]

But where string theory lacks in scientific merit, it makes up for it in revolutions. String theorists are always talking about all the revolutions that they have accomplished. An AAAS Science magazine article about string theory started with the revolution story:

> ASPEN, COLORADO — Twenty years ago, this chic playground for skiers and celebrities gave birth to a scientific revolution.

Greene wrote a popular string theory book and turned it into a 3-hour TV show in 2003. In the New York Times, he wrote:

> String theory continues to offer profound breadth and enormous potential. It has the capacity to complete the Einsteinian revolution and could very well be the concluding chapter in our species' age-old quest to understand the deepest workings of the cosmos.

Physicist John H. Schwarz's web site says that string theory "has not yet received the attention it deserves from historians of science", and then describes the discovery of five 10-dimensional string theories in 1984-85 as being the first superstring revolution. The second superstring revolution occurred in the 1990s when these theories were unified into an 11-dimensional theory. There is no general agreement on what the third superstring revolution is, except that there are more revolutions to come. A June 2010 Scientific American article suggests that *twistor* duality may be the latest superstring revolution.[408] In a field with no scientific accomplishments, a Kuhnian paradigm shift is the best that can be expected. Revolutions substitute for progress.

String theory has nothing to say about any real-world particle or fields, so it was hoped that the theory would say something about the aether, or to use today's preferred term, the vacuum. The aether has a form of energy. If string theory could predict the energy of the aether, it would be the theory's biggest accomplishment. Maybe then theorists could move on to trying to describe particles or fields as perturbations in that aether. But string theory is unable to say anything about the aether/vacuum. Weinberg says:

> A disappointing aspect of string theory is that it has so far failed to shed any light at all on what is probably the biggest outstanding problem in the physics of what we can actually see in nature — the failure to understand the energy of empty space, the so-called cosmological constant.[409]

String theory predicts a ridiculously large number of possible aether/vacuum states, each with different physical properties.[410] Instead of string theory telling us something about the physical world, it only tells us that just about anything is possible. Except that none of those possibilities bears any resemblance to our real world, as far as anyone knows. The cosmology of empty space seems to be explained by general relativity with dark energy, and string theory is supposed to be consistent with relativity, but attempts by string theorists to estimate the dark energy density have failed miserably.

The chief string theory guru is the Princeton physicist Edward Witten. He has gotten awards for his brilliant mathematics, but he refuses to discuss the scientific failures of string theory. Many of his mathematical conjectures have been proved by other mathematicians. He has no theory that could be confirmed by experiment. He acknowledges that string theory is not understood, and justifies it by saying that it is the prevailing paradigm. Any new theory will be considered a development of string theory.

Even if Witten does come up with a unified field theory that matches the standard model, we will have no way of knowing whether it is any better or worse than the standard model, because the standard model already agrees with experiment. The brilliant French mathematician Alain Connes has devised such a theory using what he calls *noncommutative geometry*. Even if something like string theory were to have some success, there will be no reason to think that it will be any better than the alternatives.

The string theorists give two main justifications for pursuing their work. They say that the theory is beautiful, and that it is the only game in town. But the theory is not beautiful, as it is vastly more complicated than the standard model, and it is unique only in the sense that the theory has the highest social status within the physics community.

String theory has inspired some worthwhile mathematics. Mathematicians are happy to study higher dimensional objects and other complexities that may not necessarily have any physical significance. But mathematicians prove their theorems. String theorists do not.

A new $9 billion European particle accelerator called the Large Hadron Collider (LHC) will probe higher energies than any previous experiment, and may even find the Higgs particle. It uses electromagnets to accelerate protons to nearly the speed of light, and then smashes them together in order to look for new particles in the resultant burst of energy. A *hadron* is a particle composed of quarks, like the proton. But the LHC will not test string theory. The string theorists look down on the LHC as just phenomenology. Their ideal is a top-down theory that can never be tested. A Scientific American article listed twelve events that will change everything, and predicts that there is a 50-50 chance that the LHC will discover the extra dimensions of string theory.[411] Meanwhile many string theorists are backing off such predictions, and now say that they would be surprised to see any such evidence in their lifetimes.[412]

We have much more important problems in understanding the big bang. One popular variant of the big bang theory says that the early universe went through a period of *inflation*, where the mass and energy of the universe exploded by a factor of millions or more. The regular big bang theory says that space was expanding rapidly, and the matter was expanding with it, but the total amount of matter was not changing. During inflation, the amount of matter was increasing rapidly.

The best explanation for inflation is that the aether was making a quantum leap to a lower energy level. The trouble with inflation theory is that no one knows when it started, when it ended, what caused it, or how intensely it took place. All of the cosmological models have problems like this.

Another hot area of debate among theoretical physicists is the Hawking black hole information paradox. General relativity teaches that nothing, not even light, can escape a black hole. Stephen Hawking predicts that a black hole will eventually evaporate after trillions of years. The question is whether information gets lost in the process. The trouble with this question is that nobody knows whether information gets lost in simple laboratory quantum mechanics experiments. Whether there is loss depends on the interpretation of quantum mechanics, and there is no consensus on that, and no known experiment to tell us.

Theoretical physicists seem to believe that they can settle questions like these by emulating Copernicus and Einstein. That is, if they adopt the right convictions about how the universe ought to be, formulate the right postulates, and derive the mathematical consequences, then they can revolutionize science. Hawking promotes his 2010 book by saying, "It was Einstein's dream to discover the grand design of the universe, a single theory that explains everything."[413] The book takes this dream beyond our universe, and pretends to explain others as well:

> But if in the light of recent advances we interpret Einstein's dream to be that of a unique theory that explains this and other universes, ... why M-theory? ... M-theory is the unified theory Einstein was hoping to find.[414]

M-theory is Witten's version of string theory. The New York Times book review says that Hawking is "the most revered scientist since Einstein".[415] Others have called him "the smartest man in the world", when they are not saying that about Witten. Hawking has now apparently repudiated his earlier positivist philosophy, and says:

> We seem to be at a critical point in the history of science, in which we must alter our conception of goals and of what makes a physical theory acceptable. It appears that the fundamental numbers, and even the form, of the apparent laws of nature are not demanded by logic or physical principle. ... People are still trying to decipher the nature of M-theory, but that may not be possible.[416]

Hawking's position now seems to be that the empiricism of science, the logic of philosophy, and the faith of religion can all be replaced by the nebulous and ill-defined M-theory. The theory explains nothing, but if it is a Kuhnian paradigm shift, then he can pretend that it explains everything. Apparently a revered scientist can follow the Einstein myth and redefine science as a mystical pursuit.

Science does not work by following the opinions of its revered leaders about how the universe ought to be. Or at least it hasn't in the past. Relativity was developed to understand electromagnetic experiments. A first-order approximate relativity theory was developed to explain those experiments, and then refined to explain Michelson-Morley. Heliocentrism was invented by the ancient Greeks, but no one had a convincing heliocentric model until Kepler developed one to explain Tycho's data.

String theory promoters like Susskind and Weinberg say that we need to expand our definition of science in order to make room for work on ideas with no experimental backing.

String theory is a colossal failure of top-down theoretical physics. Its creators paid little attention to experiments and tried to derive a theory of everything from postulates and thought experiments, in the tradition of Einstein. Or at least that is how they think it worked for relativity theory. In fact it never worked that way for relativity, and it sure hasn't worked out that way for string theory.

The Columbia University mathematician Peter Woit detailed many of the problems with string theory in a 2006 book titled *Not Even Wrong*. He got the title from a phrase Pauli used to deride theories that do not make useful predictions. A wrong theory will at least make some wrong predictions, but a not-even-wrong theory will not even do that. Although his book is an excellent scholarly account of the development of string theory, he had a hard time finding an American publisher because he was opposed by string theorists who were unhappy that he exposed weaknesses in the theory.

As Woit explains, the vacuousness of string theory is unprecedented in modern science. Never before have so many smart people expended so much effort in the name of science on a theory that is so totally untestable. There are no equations of motion or anything else that qualify as science according to traditional notions of science. Even distinguished physicists like Feynman had put it down as useless. Woit's analysis has gone unrefuted by the string theorists. They either ignored him or launched *ad hominem* attacks against him.[417] Some of them have even bragged that the theory is unfalsifiable, and suggested that science ought to be redefined to include what they are doing. To them, there was nothing new about Woit's book because string theory had always been criticized for being untestable.

In spite of the furor caused by Woit and other string theory critics, there is actually very little disagreement about what the theory has and has not accomplished. Most all of the experts agree that the theory fails to make any unambiguous testable predictions, that it fails to even explain empty space, and that there are hopelessly many variants of the theory. They pursue it because they say that it is the only popular unified theory that has not been ruled out.[418]

Woit wrote:

> No matter how things turn out, the story of superstring theory is an episode with no real parallel in the history of modern physical science. More than twenty years of intensive research by thousands of the best scientists in the world producing tens of thousands of scientific papers has not led to a single testable experimental prediction of the theory.[419]

He is right, but string theory marches on anyway. In a 2007 debate, string theorist Michael Duff argued:

> The trouble with physics, ladies and gentleman, is that Lee Smolin and Peter Woit having lost their case in the court of science, are now trying desperately to win it in the court of public opinion. Thank you.

To Duff, winning in the court of science just means that the theory is popular with elite professors, not that it has accomplished anything. In a 2002 interview, he said that the theory has two challenges — pinning down what the theory is, and making contact with experiment. And yet he slams the skeptics as pseudoscientists for not accepting the academic consensus.[420]

Duff also challenged a critic's notion that good theories are quickly validated with evidence. As a counterexample, he said that Einstein described the problem with quantum entanglement in 1935, but it was not tested until the 1980s.

The history of science is filled with examples of hypotheses that could not be tested until centuries later. The ancient Greeks had conjectures about matter being made of atoms, and about the Earth going around the Sun, and these were not resolved for millennia. Newton hypothesized that light was made of particles, not waves, and it took 20th century quantum mechanics to prove that it was both. Many other conjectures took centuries to resolve, such as the germ theory of disease. But it is much stranger to find a whole theory like string theory that is so broadly pursued without being testable at all.

Quantum entanglement was developed in the 1920s along with the basics of quantum mechanics. In any system of multiple particles, the particles are usually entangled. That means that the particles are treated jointly, and a measurement of one particle affects our predictions about the other particles. The theory was successfully applied almost immediately to problems of chemistry, where multiple electrons form a chemical bond between two atoms. What Einstein and his co-authors pointed out in 1935 was the possibility that distant particles could appear to be entangled in a quantum state, and that might be contrary to the intuitions that people had before quantum theory. Einstein hoped for a theory that was more compatible with those intuitions. It might have been a Kuhnian paradigm shift if he had persuaded others to adopt his view.

It did indeed take decades for physicists to figure out experiments to prove that Einstein's intuitions were wrong. But that was largely because there was overwhelming evidence in favor of the quantum theory, and hardly anyone saw any point to pursuing a theory of the sort that Einstein suggested. It was useful to experimentally rule out a class of possible alternate theories, but the theory of quantum mechanics had already been so broadly confirmed that no one would have known what to do if the experiments failed. String theory is not analogous at all. There is no test for string theory at all, and no experimental reason for thinking that the theory has any validity.

Greene wrote a couple of popular books on string theory, and hosted a television show on the subject. Greene answered Woit's criticism by writing a New York Times op-ed column that mentioned Einstein eleven times:

> Seventy-five years ago this month, The New York Times reported that Albert Einstein had completed his unified field theory — a theory that promised to stitch all of nature's forces into a single, tightly woven mathematical tapestry. But as had hap-

> pened before and would happen again, closer scrutiny revealed flaws that sent Einstein back to the drawing board. …
>
> String theory continues to offer profound breadth and enormous potential. It has the capacity to complete the Einsteinian revolution …
>
> But to suggest dropping research on the most promising approach to unification because the work has failed to meet an arbitrary timetable for complete success is, well, silly.

Greene's PBS Nova TV show mentioned Einstein 60 times in just the first hour. Einstein died about 25 years before string theory got started. The promise of completing the Einsteinian revolution appears to be string theory's chief selling point.

Aaron Bergman wrote the string theory response to Woit's book:

> Such a unification is called a theory of quantum gravity. The problem, however, is that this incompatibility has proven to be almost completely impenetrable to experiment. This is fairly unique in the history of physics. In this field, there have been almost no unexpected experimental results coming for three decades. …
>
> Such a situation is not entirely without precedence, however. At the turn of the twentieth century, Einstein was presented with the incompatibility of Newton's theory of gravitation and his newly developed theory of special relativity. Almost without experimental input, and with a little help from mathematicians, Einstein was able to reconcile these two theories into his theory of general relativity, a profound new understanding of the nature of space and time.
>
> The hope then is that we could, as a field, be like Einstein and solve our current conundrum by thought alone.[421]

This is really wrong. There have been many conundrums that must have seemed impenetrable to experiment. But none of them were ever solved without experimental input. For millennia, for example, it must have seemed impossible that anyone would ever figure out what could be burning within the Sun.

What is unique about string theory is not the difficulty with experiment, but with the broad adoption of a top-down theory that can never be tested. String theorists don't even seem to have much interest in any physical observations. Greene said:

> The thing that often gets lost in the discourse is that string theorists follow the mathematics — they don't dream up this or that wizardry to explain some particular thing, they follow where the mathematics takes them.[422]

In this, the string theorists are following the ancient Pythagorean cult who believed that reality is mathematical in nature at its deepest level. Wilczek wrote:

> One of the great visions of natural philosophy, going back to Pythagoras, is that the properties of the world are determined uniquely by mathematical principles. A modern version of this vision was formulated by Planck, shortly after he introduced his quantum of action. ... The ideal Pythagorean/Planckian theory would not contain any pure numbers as parameters. (Pythagoras might have excused a few small integers).[423]

String theorists also seem to emulate Galileo. He could write an insulting polemic, arrogantly declare the supremacy of mathematics for understanding the universe, and be completely wrong, all at the same time.

Einstein did not discover new physics by just following the mathematics like the string theorists. The first papers on black holes, deflection of starlight, precession of Mercury's orbit, and gravity waves were all written before Einstein ever wrote anything on gravity. The development of general relativity was directly concerned with explaining these things. It was never anything like string theory. When Einstein did try to just follow the mathematics in a search for a unified field theory, he found nothing of value.

A leading string theory physicist, Daniel Friedan, wrote this in a 2002 paper:[424]

> The long-standing crisis of string theory is its complete failure to explain or predict any large distance physics. String theory, as it stands, cannot say anything definite about large distance physics. String theory, as it stands, is incapable of determining the dimension, geometry, particle spectrum and coupling constants of macroscopic spacetime. String theory, as it stands, cannot give any definite explanations of existing knowledge of the real world and cannot make any definite predictions. The re-

> liability of string theory cannot be evaluated, much less established. String theory, as it stands, has no credibility as a candidate theory of physics.

Most of the string theory promoters are undeterred by its failures. They just double down and make even grander claims. Hawking's popular 1988 book conceded that the creation of the universe may have been an act of God that science cannot explain.[425] Now his latest best-seller says that string theory can explain it.[426] Supposedly the theory can explain the creation of our universe, as well as all of the other (unobservable) universes, without any help from God.

Woit does not quite get at the root of the problem with string theory. Yes, string theory is unscientific, speculative, nonrigorous, ugly, overhyped, and oversold. It has not made any substantial progress towards its goal of a consistent unification of gravity with the standard model. At best it is some sort of Kuhnian paradigm shift, the success of which can only be measured by how many people in the field have adopted the new view, rather than because of its scientific merit.

The problem with string theory is worse. Even it if did achieve its goal, its value would be questionable. It would replace a well-understood theory with a much more complicated one, and there would be no proof that the new theory would be any better than the old one.

String theory is also misguided. The Einstein myth says that the ideal in theoretical physics is to do thought experiments that revolutionize the subject, without needing experimental data or mathematical proof. Physics does not advance that way.

Somehow theoretical physics has been taken over by string theorists who are obsessed with completing the Einsteinian revolution. In the words of physicist Steven Weinberg, they are dreaming of a final theory.[427] They don't need any observational data. Einstein didn't need any data. Data is for phenomenologists. They don't need mathematical proofs either. Proofs are for mathematicians. Or so they say.

Sometimes physicists will argue that string theory is the "only game in town". This is just another way of saying that it is legitimate because it is the dominant paradigm. Scientific theories had always been justified by their ability to explain observations and predict experiments. But since a Kuhnian paradigm shift does not need to have any measurable or rational advantages, a paradigm can be successful just because the leaders of the field promote it.

String theorists are like the stereotypical medieval alchemists who were always talking about turning lead into gold, but never actually doing it. They are testing how long a Kuhnian paradigm shift can persist without any empirical support. As long as the leaders declare that only string theory meets their philosophical goals, then it will dominate research.

Theoretical physicists also sometimes argue that string theory must be right because it appears to be consistent. But all of mathematics is consistent, so it means nothing even if string theory turns out to be mathematically consistent. There is a lot of consistent mathematics inspired by numerology, but such consistency does not imply physical importance.

A favorite idea of the string theorists is supersymmetry. A supersymmetry is a symmetry between particles of different spins. It would mean that each boson is matched up with a similar fermion, and vice-versa. These symmetries are broken, as no known fermion has the same mass as any known boson.

A problem with this idea is that it requires one new partner particle for every known particle. None of those particles have been observed. The big hope is that the lightest such particle will be found at the LHC, and explain *dark matter*. Dark matter is an unexplained source of gravity that seems to coexist with galaxies. It was conjectured because rotating galaxies appear to violate Kepler's third law. It could be an invisible substance that drifts along with galaxies, somewhat like the way some physicists before Lorentz thought that the aether drifted along with the Earth. There are several other theoretical mysteries that might possibly be cleared up by supersymmetry. But there is no direct physical evidence for it.

Finding new symmetries in nature is usually considered *reductionist,* because a symmetry reduces the amount of work a scientist has to do to explain nature. For example, if a symmetry between two places tells you that the laws of physics are the same in both places, then an experiment in one place would have the same result in the other place. But supersymmetry is not reductionist. The standard model has about 20 parameters, but a supersymmetric version would have at least 100 extra parameters that would have to be determined by experiment. And there would be no way to do all those experiments, because the supersymmetric particles are not even detectable.

Some physicists argue that there is a precedent for a theoretical argument doubling the number of particles. They say that Dirac proposed in 1931 that all known matter particles have corresponding antimatter particles, based on his relativistic quantum theory.

It is a good story, but it didn't exactly happen that way. Pauli discovered his exclusion principle from detailed analysis of experiments. It appeared that two identical electrons could occupy a state. Others figured that if the electrons were spinning, then they would act like little magnets, and two oppositely-oriented magnetic electrons could fill an orbital. Magnetism was known to be a relativistic effect, and Dirac developed his relativistic quantum mechanics in order to explain these observed magnetic properties of electrons. His electron theory did predict properties for a positively charged particle, but Dirac was trying to explain protons. Only after others convinced him that the particle did not match the properties of a proton, did he suggest that it might be a new particle. Anti-electrons, later called positrons, were experimentally observed the next year in 1932. But even with the benefit of hindsight, Dirac's argument only suggests antiparticles for electrons and not for all particles, as is accepted today.

So far, the LHC has not found any supersymmetric particles, or any evidence for string theory or any other unified field theories. It may well turn out to be the second most famous failed experiment in history, after Michelson-Morley, and the most expensive.

Cosmology and the new orbs

Ancient cosmologists represented the sky as being composed of orbs, or celestial spheres. The Sun, Moon, and other sky objects are spherically shaped, but they were not the orbs. The orbs were giant transparent or bluish spherical shells that covered the sky and rotated around the Earth. The stars and planets were seen as being like Christmas tree lights on the orbs. That was the meaning of the term "orbs" in the title of Copernicus' book. The idea sounds fantastic and implausible, but the ancients were only concerned with explaining the appearance of the sky. They had no way of knowing whether the orbs were real or not, and had no need to speculate. Tycho's comet observations cast doubt on the orbs, because their non-circular orbits seemed to pass through the orbs. In his later writings, Kepler was the first to clearly abandon the orbs, and treat the planets as individual objects that are not attached to anything else. Kepler was concerned with reality as well as appearance.

Cosmology has made some startling progress in the last 20 years, largely as a result of advances in telescope technology. But the field is also overrun with untested hypotheses. A recent article admitted:

> The currently fashionable concordance model of cosmology (also known to the cognoscenti as "Lambda – Cold Dark Matter," or ΛCDM) has 18 parameters, 17 of which are independent. Thirteen of these parameters are well fitted to the observational data; the other four remain floating. This situation is very far from healthy. Any theory with more free parameters than relevant observations has little to recommend it.[428]

Cosmology gets much crazier than just having underdetermined parameters. A popular idea is the Boltzmann brain. The New York Times described it:

> If true, it would mean that you yourself reading this article are more likely to be some momentary fluctuation in a field of matter and energy out in space than a person with a real past born through billions of years of evolution in an orderly star-spangled cosmos. Your memories and the world you think you see around you are illusions.[429]

The search for extraterrestrial life has been unsuccessful in this universe, so some scientists are looking for life in other universes.[430] The idea is that our universe could be just one of many possible unrelated universes, and each could have its own laws of physics. The term *multiverse* is used for the collection of all universes. There are actually many different multiverse theories. Greene's latest book says that there are nine lines of reasoning that lead to defining alternate universes.[431] The *anthropic principle* says that our particular universe has been finely tuned to make life like ourselves possible.

The wishful thinking behind the multiverse is opposite that of unified field theory. While unified field theory seeks one central explanation for everything, the multiverse says that nothing can be explained because everything is possible in some universe. The universes are not even related, and may even have different laws of physics, so there is no way of knowing whether any explanation has any validity.

The futility of trying to do cosmology based on theory instead of data has been explained this way:

> If simple perfect laws uniquely rule the universe, should not pure thought be capable of uncovering this perfect set of laws without having to lean on the crutches of tediously assembled observations? True, the laws to be discovered may be perfect, but the human brain is not. Left on its own, it is prone to stray, as many past examples sadly prove. In fact, we have missed few chances to err until new data freshly gleaned from nature set us right again for the next steps. Thus pillars rather than crutches are the observations on which we base our theories; and for the theory of stellar evolution these pillars must be there before we can get far on the right track.[432]

Speculation about alternate universes has become common. A 2010 Scientific American article investigated the possibility of life in a universe without radioactive decay.[433] You might think that life without radioactivity would be easy, but the weak interaction is what keeps the Earth's interior warm and the continents drifting. It keeps the Earth geologically alive and it also has a role in the nuclear reactions that created the matter on Earth. It is hard to imagine life without it. The authors have a better imagination, I guess, because they speculate that life might still be possible with no weak interaction if various other laws of physics were changed. There is no way to know, of course.

Bruno speculated centuries ago about how Earthly events might be replayed on other worlds. He had no way of knowing whether man would eat the forbidden fruit on those other worlds, and we have no way today.

String theorist and Stanford professor Leonard Susskind has a recent book titled, *The Black Hole War: My Battle with Stephen Hawking to Make the World Safe for Quantum Mechanics*. It is 470 pages about a seemingly trivial philosophical difference. He justifies it at the beginning with:

> The Black Hole War was a genuine scientific controversy nothing like the pseudodebates over intelligent design, or the existence of global warming. Those phony arguments, cooked up by political manipulators to confuse a naive public, don't reflect any real scientific differences of opinion. By contrast, the split

> over black holes was very real. Eminent theoretical physicists could not agree on which principles of physics to trust and which to give up. Should they follow Hawking, with his conservative views of space time, or 't Hooft and myself, with our conservative views of Quantum Mechanics? Every point of view seemed to lead only to paradox and contradiction. Either space time — the stage on which the laws of nature play out — could not be what we thought it was, or the venerable principles of entropy and information were wrong. Millions of years of cognitive evolution, and a couple of hundred years of physics experience, once again had fooled us, and we found ourselves in need of new mental wiring.
>
> The Black Hole War is a celebration of the human mind and its remarkable ability to discover the laws of nature. It is an explanation of a world far more remote to our senses than Quantum Mechanics and relativity. Quantum gravity deals with objects a hundred billion billion times smaller than a proton. We have never directly experienced such small things, and we probably never will, but human ingenuity has allowed us to deduce their existence, and surprisingly, the portals into that world are objects of huge mass and size: black holes.[434]

The black hole war is nothing like the global warming debate because the latter has immediate testable predictions and public policy consequences. The climate data in the coming decades will tell us who is right and who is wrong.

Intelligent design is less clear-cut, as some hypotheses are testable and some are not. Susskind wrote his 2005 book *on The Cosmic Landscape: String Theory and the Illusion of Intelligent Design*, where he made his peculiar arguments for his own version of intelligent design.

But the black hole war is not going to be decided by any observation or other objective standard. It is only a question about the opinions that may be adopted by the "eminent theoretical physicists", whoever they are. Susskind sounds like a theologian arguing about which god to pray to, not a scientist.

He admits that we "probably never will" have any data to decide these issues, and he celebrates the ability of the human mind to ponder such esoteric issues.

An Amazon reviewer suggests that Susskind may have abandoned scientific reasoning:

> If we accept the argument that something that a falling observer (someone who cannot return nor communicate with the rest of the world) can observe is considered as a valid scientific observation, we then lose our ability to criticize people for believing that the dead go to Heaven. The dead person (one who cannot return nor communicate with the rest of the world) observes Heaven. We scientists must be very careful about our scientific reasoning, and not give others the opportunity to twist it to make it sound as if we support religion, as is, unfortunately, often the case.

Susskind does not even pretend that the black hole war is a scientific question of the sort that might be settled by observation. He subscribes to paradigm shift theory, and says that "Kuhn's ideas seemed right on target".[435] He says that we are in a paradigm shift, and physicists are changing their worldview. According to Kuhn, these changes do not occur because the new view is measurably superior to the old view, or that the new view describes an objective reality any better. Instead it is just a social consensus of the leaders in the field. So Susskind is concerned with using philosophical arguments to persuade his colleagues, and with running opinions polls to see whether his view is gaining. He treats something as an established fact when it is no longer a matter of contention among his fellow theorists.

Hawking has famously proposed that some radiation may leak out of a black hole. General relativity teaches that nothing can cross the event horizon from the inside, but perhaps the aether on the event horizon might decompose into particles going in and out, with the particles going out giving the appearance of black hole radiation. The entire black hole might even evaporate away after many trillion years. The question that has since puzzled him is whether any information is lost in the process. Hawking made a wager that information was lost, and later paid up when he declared that the information was escaping into other universes.

The trouble with this so-called paradox is that there is no way that anyone can determine whether Hawking is right or wrong. Nobody even knows whether information is lost in labs right here on Earth. Quantum mechanics does a wonderful job of predicting the outcomes of certain types of experiments, but it cannot tell us whether information is lost in even the simplest experiments. The theory has multiple interpretations, and in some of them information is lost and in some it isn't. There is no experiment that distinguishes these interpretations. And there is certainly no experiment to measure information being released from black hole evaporation.

The Schroedinger cat experiment puts a cat in a closed box and then some quantum process possibly kills the cat. The scientist does not know for sure whether the cat is alive or dead until he opens the box. The possibility of being alive or dead can be thought of as a bit of information. Opening the box seems to eliminate one of the possibilities, and to destroy that information. But under some interpretations of quantum mechanics, seeing a live cat sends a dead cat to another universe, and no information is destroyed. There is no experiment that can detect which interpretation is correct.

Einstein was unhappy with such probabilistic states because he was a determinist, believing that the past determines the future without any chance involved. God does not play dice, he said. He had his own version of the Schroedinger cat example, and argued that quantum mechanics is not acceptably realistic. Susskind is sort of an inverse determinist, believing that the future determines the past. If information disappears into a black hole, then he worries that the future will not determine the past.

Tracking the information in black holes is far more difficult. No one can measure the information going in, and no one can wait the many trillions of years for a black hole to evaporate.[436]

This is not science. Maybe it is science fiction, or philosophy, or something else, but it is not science if there is no way to tell whether a hypothesis is true or not. Hawking and others justify such theorizing based on completing their version of a Einsteinian revolution.[437]

The new realism

Modern physics can be confusing because it often stops short of telling us what is really going on. We have excellent theories of quarks and black holes, but these things are so far outside our personal experience that it is hard to say how real they are. The problem of deciding the reality of a scientific description is not new. The 1911 edition of the Encyclopedia Britannica raises the issue about epicycles when it defines:

> EPICYCLE [from Greek roots], in ancient astronomy, a small circle the centre of which describes a larger one. It was especially used to represent geometrically the periodic apparent retrograde motion of the outer planets, Mars, Jupiter and Saturn, which we now know to be due to the annual revolution of the earth around the sun, but which in the Ptolemaic astronomy were taken to be real.

This definition is fine right up to the last word, "real". The usage stems from an obscure philosophical prejudice about what is real and what is not. Everyone agrees that the lunar epicycles are real. Most people agree that the epicycles for Mercury and Venus are real. They only have a problem with the epicycles for the outer planets, as those orbits are a little harder to visualize.

The above definition is like defining an electron as "a negatively-charged subatomic particle which we now know to be due to the quantization of electromagnetic fields, but which in the 1913 Bohr atomic model were taken to be real." Or like defining an elliptical orbit as "an elongated circle used to represent geometrically the periodic apparent motion of the planets, which we now know to be due to the curvature of spacetime, but which in the Keplerian and Newtonian astronomy were taken to be real." Of course, all of these scientific ideas are real and legitimate, and great advances over previous knowledge. The epicycle was a particularly great one, and was one of the most important ones in all of ancient science.

The same encyclopedia has an article about the aether, without saying whether or not it is real:

> The hypothesis of an aether has been maintained by different speculators for very different reasons. To those who maintained the existence of a plenum as a philosophical principle, nature's abhorrence of a vacuum was a sufficient reason for imagining an all-surrounding aether, even though every other argument should be against it. To Descartes, who made extension the sole essential property of matter, and matter a necessary condition of extension, the bare existence of bodies apparently at a distance was a proof of the existence of a continuous medium between them. But besides these high metaphysical necessities for a medium, there were more mundane uses to be fulfilled by aethers. Aethers were invented for the planets to swim in, to constitute

> electric atmospheres and magnetic effluvia, to convey sensations from one part of our bodies to another, and so on, till all space had been filled three or four times over with aethers. It is only when we remember the extensive and mischievous influence on science which hypotheses about aethers used formerly to exercise, that we can appreciate the horror of aethers which sober-minded men had during the 18th century, and which, probably as a sort of hereditary prejudice, descended even to John Stuart Mill.

The encyclopedia also had an article on "medium", which was only about the sort of medium who communicates with dead people. It treated such stories as being credible and real.

The issue of the reality of astronomical models is much older than 1911, and much older than Copernicus. Ptolemy understood that he was describing the apparent motions of the planets, without necessarily giving a good description of what is really going on in the sky. Centuries before him, Plato and Aristotle wrestled with these same ideas. Plato believed that numbers and other mathematics were real, and Aristotle believed in the reality of his observations. The same issues haunt physicists today. Hawking's 2010 book has a chapter on reality. It describes the conflict between the realists, who believe that science describes an objective reality, and the anti-realists, who say that the world may all be just a figment of our imaginations.

Most scientists are realists, and believe that they are studying an objective reality. Philosophers are not so sure. 20th century physics intensified the issue. Before 1900, one could doubt the reality of atoms and electric fields. Physicists used the term "electron theory" for Lorentz's electromagnetism, even though they were not sure whether electron properties were directly measurable. Yes, there were very good physical theories for atoms and electrons, but no one had seen them directly. They only saw the consequences and got persuaded by mathematical theories that explained those consequences.

The theory of relativity raised the question of whether the motion of the Earth was real. Motion depends on your frame of reference. This remarkable theory centered around Lorentz transformations explained it all, and was confirmed by experiment. And yet it was hard to say that the Lorentz transformation was real. It raised the possibility that what we think is a round Earth is really squashed flat as a pancake, and we do not notice. The theory taught that the Earth looks that way from a rapidly moving frame of reference.

It is common to say that the rotation of the Earth is real, but the Coriolis force is fictional. That is, an artillery shell and a Foucault pendulum only appear to be deflected because the Earth is rotating underneath. With relativity, one can say that the artillery shell is really going in a straight line and only appears to have parabolic trajectory because spacetime is curved. One can say that all clocks really keep time at the same rate and only appear to slow down with motion or gravity because we compare them without properly considering the 4-dimensional geometry.

This view of reality is preferred by mathematicians, but it has its limits. All hurricanes rotate counterclockwise (in the northern hemisphere) because of the Coriolis effect. The storms are real. Magnetism is caused by time slowing down for the electrons in a metal. The magnets are real. The geometry gives a nice explanation, but what we observe is real.

Quantum mechanics raised questions about the reality of subatomic physics in the 1920s. No one doubted the reality of atoms anymore, but the theory could not tell us whether electrons and photons were really particles or waves. The theory could accurately predict experiments that measured useful quantities like energy and momentum, but no one could confidently tell you what was really going on.

There are about a dozen viable interpretations of the theory of quantum mechanics. They vary a lot in terms of their philosophical implications. Some are probabilistic, and some are deterministic. Some require a conscious observer making measurements, and some do not. All of these interpretations are consistent with known experiments. There is a Nobel Prize waiting for anyone who can prove that one of these interpretations is more correct than the others.

If you are looking for an explanation of electrons in terms of everyday objects like ping-pong balls, you are going to be disappointed. Whatever an electron is, it is nothing like those everyday objects. Electrons have particle and wave properties. Any explanation of them as just particles or just waves is going to be just an unsatisfying metaphor that should not be taken literally. Such explanations of quantum mechanics tend to lead people to believe that reality is very strange at the atomic level. Thus electrons are real, but there is no simple answer for what they really are.

Einstein is often praised for his supposed belief in realism. Some say that he showed that atoms really exist, that light is really photons, that the aether is not real, that spacetime is really curved, and that quantum mechanics fails to explain what is really going on inside an atom. And yet he is also praised for his willingness to ignore experiment and to argue how the world ought to be. His famous 1905 relativity paper was operationalist, and not realist, because he only tried to explain the outcome of measurement operations.

Einstein himself said confusing things about his belief in reality:[438]

> "The physical world is real."
>
> The above statement appears to me, however, to be, in itself, meaningless, as if one said: "The physical world is cock-a-doodle-do." It appears to me that the "real" is an intrinsically empty meaningless category (pigeon hole) … I concede that the natural sciences concern the "real," but I am still not a realist.

Modern theoretical physicists do not seem to be bound by traditional understandings about the limits to our knowledge. Hawking's 1988 book explained the physics of the big bang, and said that we cannot know why God created the universe. He was apparently an atheist, but, like Einstein, he was comfortable talking about God as a source of cosmic law. At that time, he was still subscribing to a positivist philosophy. In his 2010 book, he declares that philosophy is dead, and so is God. He adopts a broad new idea of reality that is no longer constrained by physical observation or mathematical proof. He claims that the new physics can tell us why the universe was created.

Some physicists go further, and do not distinguish between mathematical abstractions and reality. The Swedish-American cosmologist Max Tegmark calls this the *mathematical universe hypothesis*. It is the modern version of the ancient solipsist idea that life is just a dream. The trouble with this idea is that there is no way to test it, or even to give any meaning to it. We do not know whether our universe can be represented by a mathematical abstraction, and we do not know whether anything can be known about other universes.

Hawking says that new theories of the multiverse, supersymmetry, string theory, and M-theory explain everything, even explain those ultimate questions that were formerly reserved for God. He is vague on the details, because no one has succeeded in formulating any coherent theory for any of this yet, and there will be no way to test it, if and when someone does formulate such a theory. Nevertheless, he suggests that we believe it because the physics community declares it to be real.

Hawking ridicules Ptolemy for believing that epicycles were real, and Lorentz for believing that the aether was real. And yet he wants to convince us that superstrings and alternate universes are real. The physical theories behind the epicycle and the aether are among the most well-accepted and well-confirmed theories in all of science. These theories have no plausible alternative. Hawking's position is more nonsensical that Sokal's hoax. If Hawking were to suddenly announce that his book was a hoax and that any undergraduate physics or math major would realize that it is a spoof, then I would have to agree that he has written a better spoof than Sokal did.

Einstein ruined physics

The public is saturated with books and TV shows about Einstein, his life story, and his physics. But some mysteries remain. How did an ordinary physicist come out of nowhere to do such brilliant work on relativity, and then waste the rest of his life on such unproductive projects? How could one person be so much smarter than everyone else?

The explanation may be quite simple. He did not come out of nowhere. He had a doctoral degree in physics. His wife was a physicist. He was well-read on current research. His work on relativity was not really so brilliant or original. He convinced everyone that he was a great genius, and eventually he believed it himself. He was not the first to bamboozle the public, and he won't be the last.

It may seem unbelievable that so many people could be so wrong about Einstein. But there is actually a long history of books and articles pointing it out. The idea that Lorentz and Poincare discovered special relativity is not new. Pauli (1921,1956),[439] H. Thirring (1927), Born (1924,1962),[440] Whittaker (1953),[441] Feynman (1964), Hawking (1988),[442] Penrose (2002,2004)[443] and Wilczek (2008)[444] all acknowledge Poincare's priority, although some of them prefer to credit Einstein anyway. Research by historians in the last 20 years has clarified the matter further, and shown that Poincare had it all before Einstein.

Myths about Einstein have influenced modern physicists to have unrealistic expectations about what can be done with theoretical work. Hawking's view is probably typical of theoretical physicists today, both in terms of drawing the wrong lessons from the relativity story and in believing in ignoring experiment. He said:

> In theoretical physics, the search for logical self consistency has always been more important in making advances than experimental results. Otherwise elegant and beautiful theories have been rejected because they don't agree with observation, but I don't know of any major theory that has been advanced just on the basis of experiment. The theory always came first, put forward from the desire to have an elegant and consistent mathematical model. The theory then makes predictions, which can then be tested by observation, ...[445]

This is wrong. All the major physics theories were based on experiment. Hawking goes on with his only example, relativity:

> [Lorentz and others] introduced ad hoc postulates, such as proposing that objects got shorter when they moved at high speeds. The entire framework of physics became clumsy and ugly. Then in 1905 Einstein suggested a much more attractive viewpoint, in which time was not regarded as completely separate and on its own. Instead it was combined with space in a four dimensional object called spacetime. Einstein was driven to this idea not so much by the experimental results ...

This is an inaccurate account of relativity. Local time was invented in 1895 and Poincare understood it to be the end of separate time, even if Lorentz did not. Einstein did not combine time and space into a 4-dimensional object in 1905. He did not suggest any more attractive viewpoint than what Poincare had already presented. It was Poincare who invented spacetime in 1905, and he did it based on those experiments and ad hoc postulates that Hawking decries as ugly and unimportant. Hawking acts as if it were a good thing to theorize about what ought to be true, while ignoring previous work.

Those who credit Einstein for relativity seem to be coming from a Kuhnian mindset that denies objective reality. When they say that Einstein was the true inventor because of trivial terminological differences about time and the aether, they are not relying on any observational properties of the real world. They are arguing that he led a Kuhnian paradigm shift, and not that there was any scientific substance to his work. And those who promote a new Einsteinian revolution are also coming from a Kuhnian mindset.

It is true that Einstein was not so concerned with experimental results, but that is only because he was just writing an exposition of the theory that Lorentz and Poincare created, and they were directly inspired by electromagnetism and the Michelson-Morley experiment.

Einstein died before the word paradigm became popular, so he used the word *weltbild* instead. It is a German word meaning *worldview*. He used it often, such as in this description of his attitude towards top-down physics:

> The supreme task of the physicist is to seek those most general, elementary laws out of which the *Weltbild* can be achieved through pure deduction.[446]

Einstein would tell his story about how he single-handedly created relativity theory in order to justify doing formal theory and ignoring experiment:

> The theory of relativity is a fine example of the fundamental character of the modern development of theoretical science. The initial hypotheses become steadily more abstract and remote from experience. On the other hand, it gets nearer to the grand aim of all science, which is to cover the greatest possible number of empirical facts by logical deduction from the smallest possible number of hypotheses or axioms. Meanwhile, the train of thought leading from the axioms to the empirical facts or verifiable consequences gets steadily longer and more subtle. The theoretical scientist is compelled in an increasing degree to be guided by purely mathematical, formal considerations in his search for a theory, because the physical experience of the experimenter cannot lead him up to the regions of highest abstraction.[447]

He then goes on to describe relativity in terms of Poincare's ideas of imaginary time and covariance, without mentioning Poincare. If relativity is such a fine example of scientific progress, then the story should be told accurately. It is a historical fact that those grand aims and higher abstractions were not originated by Einstein, and not developed in the way that Einstein describes.

After Einstein was famous, he defended the idea that science need not be empirical. He explained himself in 1930 to the New York Times by saying:

> I assert that the cosmic religious experience is the strongest and noblest driving force behind scientific research.[448]

The newspaper, in turn, promoted the idea that his relativity was "so profound that only twelve men in the entire world were believed able to fathom its depths." Einstein was the new Messiah, and he even had his twelve disciples.

From this Einstein myth, Hawking and other theoretical physicists have learned that the best physics theories are obtained by applying abstract principles, ignoring observation, and trying to do top-down physics. This attitude is evident in the work they do. Hawking's own favorite research topic is the black hole information paradox, where there is no possibility of any experimental evidence.

This attitude is ruining physics. Theoretical physicists are chasing Kuhnian paradigm shifts instead of trying to explain observational data and objective reality. Born wrote many friendly letters to Einstein, and one exemplified the elitist attitude of many theoretical physicists:

> It is my belief that when average people try to get hold of the laws of nature by thinking alone, the result is pure rubbish.[449]

Born praised Einstein for such thinking, and called it "Jewish physics". Einstein accepted that as a compliment.

The Einstein myth persists. Every theoretical physicist wants to concoct some new principle, do some thought experiments, derive some equations, and revolutionize physics just like Einstein. Somebody should tell them that Einstein himself never did that, and the result was pure rubbish whenever he tried.

The American string theorist David Gross says that Einstein is his hero and model, and says that he was inspired by this favorite Einstein quote:

> The supreme test of the physicist is to arrive at those universal laws of nature from which the cosmos can be built up by pure deduction.[450]

Gross goes on to say that space and time will be doomed in the next string theory revolution.

The goal of completing the Einstein revolution is not a worthwhile goal. Arguments are not really scientific unless there is some way to tell whether they are right or wrong. That means using experiments or observations.

Mathematical physics (in the sense of proving theorems) is not the Einsteinian revolution. Einstein did not care about mathematically rigorous theorems, and his modern-day emulators do not either. String theory has inspired some outstanding mathematical physics. But string theory itself is not mathematics. The purpose of string theory is to create an Einsteinian revolution, not to prove mathematical theorems.

Einstein is the new Aristotle. Physicists love to ridicule Aristotle for his non-quantitative theory of physics, for his thought experiments, for his unsubstantiated realism, and for his (supposed) attempts to explain the world according to how he thought the world ought to be, instead of how it is. Most of all, they ridicule Aristotle's followers for idolizing the master, and for blindly following what he had to say.

Aristotle was a great genius. His reasoning was influential for well over a millennium. But Einstein's fame is based on the work of others, and his legacy is the pursuit of unscientific dreams. Now he is idolized more than Aristotle ever was, and his followers have created a subject more sterile than millennium-old Aristotelian physics.

Medieval monks are mocked for debating how many angels can dance on the head of a pin. They didn't really do that, but modern theoretical physicists write papers on topics nearly as silly. They write papers on alternate universes, black hole information loss, extra dimensions, and Boltzmann brains. Most of them are preoccupied with string theory, which has no connection to the real world. And they all say that they are pursuing Einstein's dreams.

It is time to stop idolizing Einstein. More importantly, it is time to stop pretending that physics needs some sort of Einstein-style revolution in order to promote some philosophical view that no one would be able to test. It is time to start recognizing scientific progress by how it resolves hypotheses about the observable world.

7. Endnotes

Additional notes and links are at www.DarkBuzz.com.

[1] Gallup poll, December 31, 1999.
[2] Physics World, Nov. 1999.
[3] As counted by John Horgan, NY Times, 2006.
[4] Colbert Report, TV show on the Comedy Channel, Oct. 28, 2009.
[5] Overbye calls it "arguably called the most famous scientific paper in history." 2000 book, p.135.
[6] Others had expressed similar concepts. Lagrange said, "One may view mechanics as a geometry of four dimensions" in about 1788. A French mathematician, d'Alembert, said similar things in 1754.
[7] Einstein's 1905 paper did mention Lorentz by name in connection with his electromagnetism, but not his relativity or his transformations.
[8] E.g., Olivier Darrigol, The Genesis of the theory of relativity, Séminaire Poincaré 1: 1-22, 2005. "The empirical equivalence of the two theories simply results from the fact that any valid reasoning of Einstein's theory can be translated into a valid reasoning of Poincaré's theory by arbitrarily calling the time, space, and fields measured in one given frame the true ones, and calling all other determinations apparent."
[9] Einstein, Why Socialism?, Monthly Review, May 1949.
[10] The FBI said that Einstein belonged to 34 Communist front organizations. He also was a Stalin apologist, as documented in The Myth of Consistent Skepticism: The Cautionary Case of Albert Einstein, by Todd C. Riniolo and Lee Nisbet. Skeptical Inquirer, Volume 31.3, May-June 2007. Such people are known as communist fellow travelers.
[11] The Private Lives of Albert Einstein, by Roger Highfield and Paul Carter. St. Martin's Griffin (March 15, 1994)
[12] In 1954 Einstein wrote, "The word god is for me nothing more than the expression and product of human weakness, the Bible a collection of honorable, but still primitive legends which are nevertheless pretty childish."
[13] The World Set Free, H.G. Wells, 1914.

[14] Planck used the letter *h* for his proposed new constant. It is sometimes more convenient to divide by 2π, and that is how *h-bar* is defined. Both are called Planck's constant, with the latter being sometimes called the *reduced* Planck's constant.
[15] Thematic origins of scientific thought: Kepler to Einstein, by Gerald James Holton, p.191.
[16] Poincare, St. Louis lecture, 1904.
[17] Hertz, Heinrich (1892) Untersuchungen über die Ausbreitung der elektrischen Kraft, Leipzig: Johann Ambrosius Barth. Electric Waves, New York: Dover 1962.
[18] Maxwell, A treatise on electricity and magnetism, Oxford, 1873. Sections 600, 601.
[19] Einstein, Elementary derivation of the equivalence of mass and energy, 1935.
[20] Poincaré, A propos de la théorie de M. Larmor, L'Eclairage électrique 5 (October 5, 1895), 5–14; reprinted in Œuvres de Henri Poincaré, Vol. 9 (Paris: Gauthier-Villars, 1954), pp. 395–426, as quoted by Shaul Katzir, 2005.
[21] Einstein's own statement of the relativity principle in 1948 was nearly identical to Poincare's.
[22] Pais, 1982. Einstein said this in 1907.
[23] Lorentz, The Relative Motion of the Earth and the Aether (1892).
[24] From a dialogue based on Galison's book Einstein's Clocks, Poincaré's Maps (2003).
[25] Einstein's clocks, Poincaré's maps: empires of time, by Peter Galison. W. W. Norton, 2004, p.219. It quotes Poincare's Science and Method, chapter on Mechanics and Optics, 1908.
[26] 7.3 Billion Years Later, Einstein's Theory Prevails, NY Times, Oct. 29, 2009.
[27] Nature podcast, Oct. 29, 2009, at 16:42.
[28] Paul Drude used *c* for light speed in 1894.
[29] Poincare, St. Louis, 1904.
[30] The German astronomer Simon Marius also discovered the moons, and named them.
[31] Spyros Sakellariadis. Descartes' experimental proof of the infinite velocity of light and Huygens' rejoinder. Archive for History of Exact Sciences, 26:1 - 12, 1982.
[32] Poincare, 1905 short paper. Miller [1994, p.81] mistranslates this section to say that Poincare's goal was a theory in which everything in the universe is of electromagnetic origin.
[33] Poincare, 1905, long paper.
[34] Today, the phrase "imaginary time" is used for the reverse concept -- getting a Euclidean metric from a spacetime metric.
[35] Einstein, Relativity, The Special and General Theory, chap. 17.

[36] Weyl, Space Time Matter, 1922 translation, p.173. Weyl confusingly attributes the solution to Einstein in 1905, and then says that it was discovered by Minkowski in 1908.
[37] As quoted by Kevin Brown, mathpages.com.
[38] Einstein complete works, edited by Stachel et al. 1989, p. 253.
[39] Einstein's Clocks: The Place of Time, by Peter Galison. The University of Chicago Press, 2000, p.355.
[40] Einstein letter to his friend Conrad Habicht, 1905. The same letter said that he had just written a "very revolutionary" paper on the energy properties of light.
[41] Poincare, 1905, long paper, end of introduction.
[42] FitzGerald, The Ether and the Earth's Atmosphere, AAAS Science 13: 390, 1889.
[43] Lorentz, The relative motion of the earth and the ether. Amsterdam, 1892. (in Dutch)
[44] Subtle Is the Lord: The Science and the Life of Albert Einstein, Abraham Pais. Oxford Univ Press, 1982, p.141.
[45] Harvey R. Brown, The origins of length contraction: I. The FitzGerald-Lorentz deformation hypothesis, 2001. Brown cites a 1915 Lorentz letter to Einstein.
[46] Lorentz 1895 discusses Michelson-Morley in the Introduction and Section VI.
[47] This assumption is explicit in Lorentz's 1904 paper.
[48] Joseph Larmor, On the ascertained Absence of Effects of Motion through the Aether, in relation to the Constitution of Matter, and on the FitzGerald-Lorentz Hypothesis (1904). Henry Andrews Bumstead, Applications of the Lorentz-FitzGerald Hypothesis to Dynamical and Gravitational Problems (1908).
[49] Einstein, On a Heuristic Point of View about the Creation and Conversion of Light, 1905, first sentence.
[50] Textbook by August Foppl, as cited by John Stachel.
[51] Poincare, ICM 1897, translated in 1898.
[52] Maxwell, Matter and Motion, 1877. The 1920 edition had notes by Joseph Larmor. Sect. 102.
[53] Statement to the Cleveland Physical Society, 1952, honoring Michelson.
[54] Weyl, Philosophy of mathematics and Natural Science, Atheneum, New York, 1960. p.96.
[55] Weyl, Space Time Matter, 1922 translation, p.173.
[56] Minkowski, The Fundamental Equations for Electromagnetic Processes in Moving Bodies, 1908. Also delivered as a conference lecture.
[57] Mathpages.com
[58] Joseph Polchinski, All Strung Out. American Scientist, Jan-Feb 2007.

[59] Felix Klein, A comparative review of recent researches in geometry, 1872. It is known as the Erlangen program.
[60] Poincare, 1907. He also said that the translation would be "very difficult and produce few benefits."
[61] Minkowski (1907/1915). The principle of relativity, Annalen der Physik 352 (15): 927–938.
[62] Credit to Poincare for 4-dimensional spacetime is in R. Marcolongo, Sugli integrali dell'equazione dell'elettrodinamica, Rendiconti della Regia Accademia dei Lincei, s. 5, v. 15 (I sem. 1906), pp. 344-349.
[63] As quoted in Scott Walter, Minkowski, Mathematicians, and the Mathematical Theory of Relativity, 1999.
[64] Quoted in Einstein: His Life and Times, by Philipp Frank, p.206.
[65] Poincare, 1902, chap. 6, The Classical Mechanics.
[66] Some of the defects in Einstein's argument are explained by Márton Gömöri and László E. Szabó, Is the relativity principle consistent with electrodynamics?, arXiv:0912.4388v2, 2009.
[67] Minkowski, The Fundamental Equations for Electromagnetic Processes in Moving Bodies (1908).
[68] Freeman J. Dyson, Clockwork Science. The New York Review of Books, vol. 50, num. 17
[69] Science Historian at Work: Peter Galison; The Clocks That Shaped Einstein's Leap in Time, NY Times, June 24, 2003.
[70] Lubos Motl, The Reference Frame, blog post on Verlinde.
[71] H. A. Lorentz, The Relativity Principle, three lectures, 1914.
[72] Poincare's 1888 lectures were published as an 1890 book, Electricity and Optics. Later lectues on electromagnetism were published in 1904 under the title, Maxwell's Theory and Hertzian Oscillations.
[73] Poincare, The Value of Science, 1905.
[74] Maxwell, Ether, Encyclopaedia Britannica Ninth Edition 8: 568–572, 1878.
[75] Lorentz, 1904, sections 1, 3.
[76] From a tribute on the Trunity College Dublin web site. The Feb. 22, 1900 Dublin address was published in the Journal of the Institution of Electrical Engineers, vol. XXIX, 1899-1900.
[77] Paul Drude, The Theory of Optics, 1902 translation.
[78] Larmor, Aether and Matter, 1900. Footnote on p.334.
[79] The quote is from The Structure of the Aether (1907), Nature, 76 (No. 1966): 222. The credits are from On the Electromagnetic Mass of a Moving Electron, Philosophical Magazine, 1907, 6 14 (82): 538-547.
[80] Ether and the Theory of Relativity, 1922.
[81] Einstein, Concerning the Aether, 1924.

[82] Einstein, "Letter to H. A. Lorentz, 15 November 1919," EA 16 494. Einstein's use of the aether after 1916 is documented in Einstein and the Ether, by Ludwik Kostro. C. Roy Keys Inc, Montreal Apeiron, 2000.
[83] Einstein, A Brief Outline of the Development of the Theory of Relativity. Nature, 106 (No. 2677); 1921. p.782-784.
[84] Einstein in Mein Weltbild, Amsterdam, Querido, 1934.
[85] Einstein and Infeld, The Evolution of Physics. Simon & Schuster, New York, 1938, p.185.
[86] Einstein, Foreword to Concepts of Space, by Max Jammer, Harvard University Press, 1954.
[87] W. Pauli, p.4.
[88] Paul Dirac, Is there an Aether?, Nature 168 (1951), p. 906.
[89] Edmund Whittaker, A History of the Theories of Aether and Electricity, 2nd ed, 1953, preface.
[90] Frank Wilczek, Physics Today (January 1999).
[91] Fantastic Realities: 49 Mind Journeys And a Trip to Stockholm, by Frank Wilczek World Scientific Publishing Company, 2006, p.278.
[92] Frank Wilczek, 2005 lecture, MIT.
[93] Frank Wilczek, Nature, vol 435, 12 May 2005.
[94] The Lightness of Being: Mass, Ether, and the Unification of Forces, by Frank Wilczek. Basic Books, 2008. P.74 and Glossary, p.228.
[95] Frank Wilczek, Quantum Field Theory, arXiv:hep-th/9803075v2, 1998.
[96] Understanding Media: The Extensions of Man, by Marshall McLuhan. McGraw Hill, NY, 1964
[97] George F. Smoot III, Cosmic Microwave Background Radiation Anisotropies: their discovery and utilization. Nobel lecture, December 8, 2006.
[98] Mathpages.com
[99] Isaacson, p.502-503. The spy was Margarita Konenkova, and the affair lasted from 1941 to 1945.
[100] New York Times, Aug. 7, 1945, p.1.
[101] Time magazine, July 1, 1946.
[102] Einstein badmouthed Poincare at a 1911 conference, cited him for non-Euclidean geometry in 1921, and denied in 1953 that he had known about Poincare's relativity papers. Einstein even denied to biographer Abraham Pais that he ever read Poincare's long 1905 relativity paper. But Einstein had mentioned ideas from that paper in a 1919 letter to Hilbert.
[103] Einstein's $E = mc^2$ mistakes, Hans C. Ohanian, arXiv:0805.1400v2, 2008.
[104] Planck, The Principle of Relativity and the Fundamental Equations of Mechanics, presented March 23, 1906.
[105] For an explanation, see A.A. Logunov, Henri Poincare and Relativity Theory, 2005, sect. 9.
[106] Minkowski, The Fundamental Equations for Electromagnetic Processes in Moving Bodies, 1908, translated in 1920.

[107] Lorentz, Columbia University lectures, 1909.
[108] Poincare, Science and Hypothesis, 1902, chap. 7.
[109] This is detailed in chap. 5.
[110] Einstein, 1905 relativity paper, end of section 4.
[111] Bernstein, Cranks, Quarks, and the Cosmos, p.22. See also Relativistic Behaviour of Moving Terrestrial Clocks, by J. C. Hafele, Nature 227, 270 - 271 (18 July 1970).
[112] Einstein, About the relativity principle and the conclusions drawn, 1907, part I, section 4.
[113] According to an S. Weinberg essay, "Edmund Halley, Christopher Wren, and Robert Hooke all used Kepler's relation between the squares of the periods and the cubes of the diameters (taking the orbits as circles) to deduce an inverse square law of gravitation, and then Newton extended the argument to elliptical orbits."
[114] Poincare, Science and Hypothesis, 1902. Chap. 10.
[115] Maxwell's Theory and Wireless Telegraphy, by H. Poincare and Frederick K. Vreeland. McGraw Publishing, New York, 1904. Poincare wrote the part on Maxwell's theory, and Vreeland translated that part to English and added a part on Wireless Telegraphy.
[116] Ibid, chap. VII, p.71.
[117] De Sitter, Monthly notices of Roy. Astr. Soc., March, 1911, p.388. As cited by Cunningham, 1915, p.83.
[118] Einstein wrote his paper on the precession of Mercury's orbit shortly before knowing the field equations that he and Hilbert published in 1916, so he used Grossmann's 1913 field equations and some calculations by Besso. They were equivalent for the purposes of solar system calculations.
[119] Einstein, Relativity: The Special and General Theory, chap. 16.
[120] Harvey Brown 2005 argues that Einstein never adopted a purely kinematical special relativity.
[121] Pauli, 1981, as quoted by Brown.
[122] This is some support for this view in the writings of Weyl, Pauli, Eddington, and others. See Harvey R. Brown, 2005.
[123] Minkowski, Space and Time (1908), sect. II.
[124] Miller, Why did Poincaré not formulate special relativity in 1905?, in Jean-Louis Greffe, Gerhard Heinzmann, Kuno Lorenz, Henri Poincaré : science et philosophie, Berlin, 1994, p.69–99.
[125] Isaacson, Einstein: His Life and Universe, 2007, p. 134-135. Says "Even more surprising, and revealing, is the fact that Lorentz and Poincaré never were able to make Einstein's leap even after they read his paper. Lorentz still clung to the existence of the ether … For his part, Poincaré seems never to have fully understood Einstein's breakthrough." Also quoted with approval by Roger Cerf, Am. J. Phys., Vol. 74, No. 9, September 2006. But Holton, an-

other Einstein biographer, says "Poincaré had understood Einstein's message only too well." Thematic origins of scientific thought: Kepler to Einstein, p.202.
[126] Pais, 1982, p.145.
[127] Galison, Einstein's Time, 2003.
[128] Hawking-Mlodinow, The Grand Design, 2010, p.96-97.
[129] E.g., Kaufmann's papers said that Lorentz's and Einstein's theories were formally identical.
[130] Einstein in Love: (A Scientific Romance), by Dennis Overbye, 2000. p.138.
[131] Poincare, The Relativity of Space, from Science & Method.
[132] Poincare, Space and Time, published in Last Essays, 1913, p.23.
[133] Tilman Sauer, The Einstein Varicak Correspondence on Relativistic Rigid Rotation, arXiv:0704.0962v1, 2007.
[134] London Times, November 7, 1919.
[135] Michio Kaku, Einstein's Cosmos, WW Norton, 2004, p.118.
[136] NY Times, May 16, 1940.
[137] NY Times, April 3, 1921.
[138] Einstein's sceptics: Who were the relativity deniers?, by Milena Wazeck, New Scientist, Nov. 18, 2010.
[139] December 31, 1922 NY Times book review of textbook skeptical of relativity. It said that 98% of the books on the subject had accepted relativity, and "We say it with regret, but it is none the less true, that mathematicians and metaphysicians in Europe and America went over almost solidly to the new cult."
[140] Feb. 19, 1979 Time Cover: The Year of Dr. Einstein. It was quoting Cambridge University's Martin Rees.
[141] Valentine Bargmann essay, published in Ideas and Opinions, by Albert Einstein. Originally published by Crown in 1954.
[142] Ohanian, p.330.
[143] An Eddington letter to Einstein said, referring to the 1919 eclipse announcement, "For scientific relations between England and Germany this is the best thing that could have happened."
[144] Eddington, The Mathematical Theory of Relativity, 2nd ed. Cambridge, 1937, p.18.
[145] Whittaker, 2. Edition: A History of the theories of aether and electricity, vol. 1: The classical theories, vol. 2: The modern theories 1900-1926, London: Nelson, 1953.
[146] Born wrote, "The reasoning used by Poincaré was just the same as that which Einstein introduced in his paper of 1905 … Does this mean that Poincaré knew all this before Einstein? It is possible … [Einstein's paper] gives you the impression of quite a new venture. But that is, of course, as I have tried to explain, not true." in an essay in the book Physics in my Generation, 1969.

[147] Space-Time Cubism, NY Times, May 06, 2001, review of Miller's book on Einstein and Picasso.
[148] Miller, Why did Poincaré not formulate special relativity in 1905?, in Jean-Louis Greffe, Gerhard Heinzmann, Kuno Lorenz, Henri Poincaré : science et philosophie, Berlin, 1994, p.69–99.
[149] Thematic origins of scientific thought: Kepler to Einstein, by Gerald James Holton, p202-7.
[150] The Poincare group includes the Lorentz group and translations. The name came from E. Wigner much later.
[151] Posted under the pseudonym Bee on the Cosmic Variance blog.
[152] Lorentz, The Theory of Electrons (1909).
[153] The prize was the Bolyai Prize, and was awarded every five years. Hilbert received the 1910 prize. As quoted in Thematic origins of scientific thought: Kepler to Einstein by Gerald James Holton, Harvard University Press; Second Edition, 1988, p.202-3.
[154] Poincare, Science and Method, 1908. Reprinted in a popular 1913 book.
[155] Chaos: Making a New Science, by James Gleick. Viking, 1987. It was a finalist for the National Book Award and the Pulitzer Prize in 1987.
[156] Nassim Nicholas Taleb. The Black Swan, 2007.
[157] Dana Mackenzie, BREAKTHROUGH OF THE YEAR. The Poincaré Conjecture—Proved. AAAS Science 314 (5807): 1848-1849, 2006.
[158] Shing-Tung Yau (with Steve Nadis), The Shape of Inner Space: String Theory and the Geometry of the Universe's Hidden Dimensions, Basic Books, 2010, p.46.
[159] Tribute by Gaston Darboux, read at the annual December 15, 1913 meeting, published in Proceedings of the Academy of Sciences, 1913.
[160] Morse, F., Journal: Publications of the Astronomical Society of the Pacific, Vol. 23, No. 136, p.73
[161] Poincare's Prize, by George G. Szpiro. Penguin, 2007. The book says on p.50 that the consensus is that Poincare "was not audacious enough".
[162] Ibid. Chapter 4 tells this story in detail.
[163] Nunquam praescriptos transibunt sidera fines.
[164] Poincare, The Theory of Lorentz and The Principle of Reaction, 1900.
[165] Lorentz, Electromagnetic phenomena in a system moving with any velocity smaller than that of light. Proceedings of the Royal Netherlands Academy of Arts and Sciences, 1904, 6: 809–831.
[166] Einstein, About the relativity principle and the conclusions drawn, 1907, p.412-413.
[167] The telegram said, "As I have already informed you by telegram, in its meeting held yesterday the Royal Academy of Sciences decided to award you last year's [1921] Nobel Prize for physics, in consideration of your work on theoretical physics and in particular for your discovery of the law of the

photoelectric effect, but without taking into account the value which will be accorded your relativity and gravitation theories after these are confirmed in the future." As quoted by Abraham Pais, 1983, p.503.

[168] Jean Mawhin (October 2005), Henri Poincaré. A Life in the Service of Science, Notices of the AMS 52 (9): 1036–1044. According to Mawhin, Poincare is still the most nominated physicist, with 49 proposals between 1901 and 1912.

[169] Nobel physics prizes were given for work related to the standard model in 1965, 1968, 1969, 1976, 1979, 1980, 1984, 1988, 1990, 1995, 1999, 2004, and 2008.

[170] 14 Dec. 1922 Kyoto lecture, reported in Physics Today, Vol. 35, No.8, pp. 45-47, Aug. 1982.

[171] Peter L. Galison, Einstein's Clocks, in Science and Society: The History of Modern Physical Sciences in the Twentieth Century, Routledge, 2001. P.204.

[172] Einstein, The Development of Our Views on the Composition and Essence of Radiation, 1909.

[173] Michel Janssen, Drawing the line between kinematics and dynamics in special relativity.

[174] Lorentz, Draft of a theory of electrical and optical phenomena in moving bodies. Leiden: E. J. Brill, 1895. (in German) See especially the end of section 1 of the introduction.

[175] Some have claimed that Einstein read and made use of that 1904 Lorentz paper, such as Tony Rothman, Lost in Einstein's Shadow. American Scientist 94 (2): 112, March-April 2006.

[176] Publisher: Norton (September 8, 2008). See also Did Einstein prove $E = mc^2$?. Hans C. Ohanian, Studies in History and Philosophy of Modern Physics 40 (2009) 167–173. This says that Laue published the first valid proof in 1911.

[177] For an explanation of some of the ambiguities, see Kinematic subtleties in Einstein's first derivation of the Lorentz transformations, by Alberto A. Martinez, Am. J. Phys., Vol. 72, No. 5, May 2004.

[178] In a 1921 unpublished manuscript, Einstein lists his hidden assumptions as space being homogeneous and isotropic, and objects being independent of their history.

[179] Penrose wrote that Einstein did not appreciate the significance of the four dimensional space time picture in 1908, and did not take it seriously for about two years afterwards. See his essay, The Rediscovery of Gravity, p.59, published in It Must Be Beautiful: Great Equations of Modern Science, edited by Graham Farmelo. Granta Books, 2002.

[180] Letter to Ehrenfest, 1913.

[181] Others saw these consequences as early as 1916.

[182] Quoted by Isaacson, p.251.

[183] This story is told by Ohanian, in Einstein's Mistakes.

[184] Professor Albert Einstein, Father of nuclear physics, The Times (London), 1955.
[185] A. Einstein, B. Podolsky, and N. Rosen, Can quantum-mechanical description of physical reality be considered complete? Phys. Rev. 47 777 (1935).
[186] Freeman Dyson, Birds and Frogs, AMS Notices, Feb. 2009.
[187] Quoted in Hilbert, by Constance Reid, p.142.
[188] This is the only known example of an Einstein paper being peer-reviewed. He eventually published the paper with the referee's ideas incorporated, as someone eventually persuaded Einstein that the referee was correct.
[189] Norbert Straumann, Gauge Principle and QED. arXiv:hep-ph/0509116v1, 2005.
[190] Quoted by Isaacson, p.514.
[191] Proceedings of the Royal Society of London. Series A, Vol. 123, No. 792, 6 April 1929.
[192] Vitaly L. Ginzburg, The physics of a lifetime: Reflections on the Problems and Personalities of 20th Century Physics. Springer, 2001.
[193] Poincare, 1905 long paper, sect. 1.
[194] The Character of Physical Law, 1994. Video of the original 1964 lectures are available online.
[195] Ginzburg, p.233, 239.
[196] Michel Paty, Physical Geometry and Special relativity: Einstein and Poincaré, 1992, p.17, in Einstein et Poincaré, 1830-1930 : a century of geometry (Berlin, 1992), 127-149.
[197] Einstein, 1905, end of sec. 6.
[198] Harvey R. Brown, Physical relativity: space-time structure from a dynamical perspective. Oxford University Press, 2006. P.62-63.
[199] These Poincare letters are online.
[200] Once upon Einstein, by Thibault Damour. A K Peters, Ltd, 2006.
[201] St. Louis lecture, 1904.
[202] Pauli was only 21 years old at the time, so it is fair to assume that the opinions in the book reflected those of Sommerfeld and his other professors.
[203] The Theory of Relativity and Science, by W. Pauli, 1956. As reprinted in Writings on physics and philosophy, by Wolfgang Pauli, Charles Paul Enz, K. v Meyenn.
[204] Rigden, Einstein 1905: The Standard of Greatness, p.95.
[205] International Congress of Mathematicians, Zurich, 1897.
[206] Rigden, p.16.
[207] Attributed to Joe Theisman, former NFL football quarterback and sports analyst.
[208] Rigden, p.83.
[209] Special Relativity: A Centenary Perspective, by Clifford M. Will, 2005.

[210] From the Presentation Speech by Professor Ivar Waller. The idea of quarks was also proposed by a couple of other physicists.
[211] Einstein, Note to the note by Mr Paul Ehrenfest: "The translation of deformable surface electrons and the sentence", Annalen der Physik 328 (7), 371–384, 1907.
[212] Lorentz, The theory of electrons and its applications to the phenomena of light and radiant heat; a course of lectures delivered in Columbia University, New York, in March and April 1906, New York. Columbia University Press, 1909. sec.189, p.223.
[213] May 1905 letter to Lorentz.
[214] Jules Henri Poincaré, 2010 online edition, Encyclopaedia Britannica.
[215] The Emperor's New Mind. Concerning Computers, Minds, and the Laws of Physics, by Roger Penrose. Oxford University Press, 2002, p.240.
[216] Einstein: His Life and Universe, by Walter Isaacson. Simon & Schuster, 2007, p.134.
[217] Dyson interview in 2000, as reported in The Einstein File, by Fred Jerome. St. Martin's Press, 2002, p.xiv.
[218] Hawking, 1988, chap. 2, p.20, 28. The 2005 edition added that Poincare "to his dying day did not accept Einstein's interpretation of the theory." P.32.
[219] Gingras, The Collective Construction of Scientific Memory: The Einstein-Poincaré Connection and Its Discontents, 1905-2005. History of Science, vol. 46, no 1, mars 2008, pp. 75-114.
[220] Darrigol, The Mystery of the Einstein-Poincaré Connection, Isis 95 (4): 614–626, 2004.
[221] The journal was the official publication of the Mathematical Circle of Palermo, an Italian mathematical society based in Sicily. It was the largest mathematical society in the world at the time. See The Poincare Conjecture, by Donal O'Shea, Walker Publishing, New York, 2007, p.151.
[222] Darrigol, 2004.
[223] Einstein did mention one of Poincare's papers two months before he died, but only to downplay its significance.
[224] Gerald James Holton, Thematic origins of scientific thought: Kepler to Einstein. Harvard, 1988, p.328. He cites Prof. Oliver Lodge as saying the quote in public on 27 May 1892.
[225] Poincare, 1906, sect. 9.
[226] Roger Cerf, Dismissing renewed attempts to deny Einstein the discovery of special relativity. Am. J. Phys., Vol. 74, No. 9, September 2006.
[227] Introduction to the special theory of Relativity, Claude Kacser. Prentice-Hall, 1967.
[228] The principle of relativity and the basic equations of mechanics, Max Planck, 1906, Negotiations German Physical Society. 8, pp. 136-141 (Presented at the meeting of 23 March 1906.)

[229] Claude Kacser, Introduction to the special theory of relativity. Prentice-Hall, 1967. Sec. 2.7, p.23.
[230] Ibid., Chap. 1, Historical background.
[231] Stephen G. Brush, Why was Relativity Accepted?, Phys. perspect. 1 (1999) 184–214.
[232] Torretti, p.86.
[233] Einstein, 1905, sect. 9.
[234] These lectures were later published in book form.
[235] E.g., Planck's March 1906 paper on relativity starts by citing Lorentz 1904 and Einstein 1905 for the principle of relativity and (Lorentz) transformations.
[236] E. Cunningham, The principle of relativity in electrodynamics and an extension thereof, 1909.
[237] Torretti, p.87.
[238] Gilbert Newton Lewis and Richard Chace Tolman, The Principle of Relativity, and Non-Newtonian Mechanics. Proceedings of the American Academy of Arts and Sciences, 1909, 44: 709–726.
[239] Richard Chace Tolman, The Theory of the Relativity of Motion. Univ. of Calif. Press, 1917.
[240] Lorentz, 1909 preface to 1906 Columbia lectures, and section 194 at the end.
[241] Lorentz, Two Papers of Henri Poincaré on Mathematical Physics. Acta Mathematica. 38, pp 293–308, (written in 1914, published 1921); reprinted in Oeuvres de Henri Poincaré, Vol. XI, 247-261, 1956 (in French)
[242] Arthur I. Miller, Why did Poincaré not formulate special relativity in 1905?, in Jean-Louis Greffe, Gerhard Heinzmann, Kuno Lorenz, Henri Poincaré: science et philosophie, Berlin, 1994, p.69–99.
[243] Miller, p.70, 85, 86, 97.
[244] Rothman, Lost in Einstein's Shadow. American Scientist 94 (2): 112, March-April 2006.
[245] Einstein, 1905, end of sec. 9.
[246] Einstein, 1905, sec. 6.
[247] As translated from *sehr matt*.
[248] Minkowski, 1908, end of section II.
[249] Maxwell, A Dynamical Theory of the Electromagnetic Field, Phil. Trans. R. Soc. Lond., 1865, 155, 459-512.
[250] For an explanation of why quantum mechanics experiments have not contradicted local causality, see: Bell´s theorem and the experiments: Increasing empirical support to local realism?, by Emilio Santos arXiv:quant-ph/0410193v1, 2004. More recent research is discussed in Quantum Mechanics Braces for the Ultimate Test, by Zeeya Merali, AAAS Science, vol 331, p.1380, Mar. 18, 2011.

[251] Relativity and the lead-acid battery,
[252] Cosmos, PBS TV series, narrated by Carl Sagan.
[253] Claudius Ptolemy, Tetrabiblos. English translation by Frank Egleston Robbins, Harvard University Press, 1940. The comment about physicians is at the end of section I.2.
[254] Archimedes, The Sand-Reckoner, Third Century BC.
[255] Mathematically, an Alamgest planet orbit is an *epitrochoid*, which is defined as the difference between two parametric circles.
[256] Erasmus Reinhold was the leading mathematical astronomer in Europe, and he admired Copernicus for elimination of the equant. See The Book Nobody Read: Chasing the Revolutions of Nicolaus Copernicus, by Owen Gingerich. Penguin, 2005. p.10, 51.
[257] Martianus Capella
[258] His Place in Science by Dava Sobel.
[259] Galileo Galilei, The Assayer, in Italian, October 1623.
[260] Galileo-Kepler Correspondence, 1597. Galileo wrote, "I indeed congratulate myself on having an associate in the study of Truth who is a friend of Truth. For it is a misery that so few exist who pursue the Truth and do not pervert philosophical reason."
[261] Kepler's introduction to Astronomia Nova, as quoted in The Sleepwalkers, Arthur Koestler, Arkana Books, London, 1989, p.342-343.
[262] Kepler may have been influence by a recent book saying the Earth is a big magnet. On the Magnet, by Gilbert, 1600.
[263] De revolutionibus orbium coelestium, end of chap.10.
[264] According to the historian of science, Maurice Finocchiaro, the accusation that Galileo had intended Simplicio to be a caricature of Pope Urban VIII did not surface until 1635, some three years after his trial (Retrying Galileo, p.62).
[265] December 2006 issue, Discover Magazine. The top two books were by Darwin. Einstein's 1916 relativity book was eighth.
[266] The Literal Interpretation of Genesis, Augustine of Hippo, North Africa.
[267] Eddington published an English translation in 1931, but omitted the data.
[268] NPR Science Friday, Scientists and Advocacy?, Nov. 12, 2010.
[269] Tycho's body was exhumed in 2010 to test theories that he was poisoned.
[270] Return of the epicycle, New Scientist, 06 August 2005 by Robert Entwistle, Bracknell, Berkshire, UK
[271] The priest was Giovanni Riccioli. See The Coriolis Effect Apparently Described, arXiv:1012.3642v1, 2010.
[272] William Herschel, On the Direction and Velocity of the Motion of the Sun, and Solar System, Philosophical transactions of the Royal Society of London, Volume 95, p.233, 1805.
[273] Planck letter of recommendation to Emperor Franz Joseph, as quoted in A. Folsing, Albert Einstein, A Biography, Penguin Books, 1998. Also quoted in NY Times, Nov. 16, 1919.

[274] Weyl, Space Time Matter, 1922 translation, p.174.
[275] Ginzburg, The physics of a lifetime: Reflections on the Problems and Personalities of 20th Century Physics. Springer, 2001, p.227.
[276] This example was first published by Paul Ehrenfest in 1909, and it is known as the Ehrenfest paradox.
[277] Einstein, Dialogue on objections to the theory of relativity, 1918.
[278] Isaacson, Einstein's Final Quest, 2009.
[279] Einstein, Relativity: The Special and General Theory, 1916, chap.17.
[280] Minkowski credited Poincare in a 1907, so Minkowski was certainly knew about Poincare's spacetime and metric, but Minkowski did not credit Poincare in 1908. Sommerfeld added credits to Poincare when he reprinted the 1908 article in 1915.
[281] Einstein, Fundamental Ideas and Methods of Relativity, 1920.
[282] Poincare's Sorbonne lectures, 1906-7.
[283] Lorand von Eotvos announced his first results in 1890. The Royal Scientific Society of Goettingen (Germany) offered a prize in 1906 for improvements to his experiment.
[284] Some say that Robert Hooke had a similar law of gravity before Newton. Newton was the first to deduce Kepler's laws of astronomy from the force formula.
[285] This paper is known as the *Entwurf* paper. That is the German word for "outline".
[286] Constance Reid, Hilbert, Springer, 1970, p.105.
[287] Einstein was upset about the possibility that Hilbert might claim credit for general relativity and complained privately of "human wretchedness".
[288] L.Corry, J.Renn and J. Stachel, Belated Decision in the Hilbert-Einstein Priority Dispute, AAAS Sceince, Vol. 278, 14 November 1997. Stachel edited Einstein's collected works, and has written his own book, Einstein from 'B' to 'Z', Birkhauser 2001.
[289] F. Winterberg and Logunov et al published papers in 2004 asserting the correctness of the Hilbert draft, and refuting Corry-Renn-Stachel. The latter responded with ad hominem attacks, and then retracted them. They had no explanation for the half-page being missing without their noticing it.
A. A. Logunov et al., How were the Hilbert-Einstein equations discovered?, arXiv:physics/0405075v3, 2004.
Anticipations of Einstein in the General Theory of Relativity, by Christopher Jon Bjerknes, 2003.
Dieter W. Ebner, How Hilbert has found the Einstein equations before Einstein and forgeries of Hilbert's page proofs, arXiv:physics/0610154v1, 2006.
[290] The field equations are usually written in terms of 4-by-4 tensors, giving 16 equations. However there are symmetries that reduce them to one trace term equation and nine other independent equations.

[291] "Indeed, researchers didn't truly nail the light-bending prediction until they used quasar measurements made in the 1960s, says physicist Clifford Will of Washington University in St. Louis, an expert in tests of general relativity." SciAm.com, March 6, 2008.
[292] Pioneer and flyby anomalies.
[293] Laser Interferometer Gravitational-Wave Observatory (LIGO), arXiv:1102.3781v1, 2011.
[294] Scientific American, February 2011.
[295] John F. Donoghue, Introduction to the Effective Field Theory Description of Gravity, arXiv:gr-qc/9512024v1, 1995.
[296] Sidelights on Relativity (Geometry and Experience). P. Dutton., Co., 1923, p.28.
[297] Poincare, ICM, 1897, translated in 1898.
[298] Alain Connes, Advice to the beginner, essay from his blog. Published in The Princeton companion to mathematics, by Timothy Gowers, June Barrow-Green, Imre Leader. Princeton University Press, 2008.
[299] Poincare, ICM 1897, reworded from 1898 translation.
[300] Owen Gingerich, The Book Nobody Read. Penguin, 2004.
[301] Copernicus, De Revolutionibus, book 1, chap. 10.
[302] Galileo's Mistake: A New Look At the Epic Confrontation Between Galileo and the Church, by Wade Rowland. Arcade Publishing, 2003, p.83.
[303] LaSalle University astronomy class web site.
[304] The evolution of eusociality, Martin A. Nowak, Corina E. Tarnita & Edward O. Wilson, Nature 466, 1057–1062 (26 August 2010). Several criticisms were published, including a three-page paper with two of the pages being a listing of the 137 authors.
[305] NY Times,.Scientists Square Off on Evolutionary Value of Helping Relatives, by Carl Zimmer, Aug. 31, 2010.
[306] Simon Singh, Big Bang: The Origin of the Universe. HarperCollins Publishers, 2004, p.32.
[307] Ibid., p.46.
[308] Ibid., p.33, 332.
[309] Ibid., p.128.
[310] Singh accused the British Chiropractic Association of "happily promoting bogus treatments", without a jot of evidence.
[311] Dorothy, It's Really Oz, by Stephen Jay Gould, Time magazine, Aug. 23, 1999.
[312] Cosmic Variance blog, Feb. 2011.
[313] Lis Brack-Bernsen, Matthias Brack, Analyzing shell structure from Babylonian and modern times, 10th Nuclear PhysicsWorkshop, arXiv:physics/0310126v2, 2003.
[314] Gould, Chapter 4: The late birth of a flat earth in Dinosaur in a Haystack, Harmony Books, New York, p. 41, 1995.

[315] Jeffrey Burton Russell, Inventing the Flat Earth: Columbus and Modern Historians. New York: Praeger, 1991. Greenwood Press; 2nd edition 1997.
[316] The story has been told by philosophers John Locke (1690), David Hume (1779), Joseph Frederick Berg (1854), William James (ca 1900), and Bertrand Russell (1927).
[317] Garry Wills, A Country Ruled by Faith, NY Review of Books, November 16, 2006.
[318] Gallup News Service, July 6, 1999.
[319] National Science Foundation, Science and Engineering Indicators 2010, Fig. 7-11.
[320] Planck, General Dynamics. Principle of Relativity. Eighth Lecture, Columbia University Press, New York, 1909, translated in 1915.
[321] According to Pierre Duhem, a textbook by the priest Domingo deSoto had the law of free fall 75 years earlier.
[322] The Complexities of Being on a Roll, NY Times, Aug. 3, 2010.
[323] Hawking, 1988, p.179.
[324] Google Books Ngram Viewer.
[325] Frederick Crews points out that "Freud was unable to document a single unambiguously efficacious treatment".
[326] Isaac Asimov Memorial Debate, American Museum of Natural History, 2011, at 01:19:30.
[327] Freud's theory of the unconscious was consistent with conventional wisdom of the day.
[328] Who did most to knock man off his pedestal?, by Michael Brooks New Scientist, 20 December 2008.
[329] C-SPAN2 Book-TV interview, March 27, 2011. The Hidden Reality, 2011, p.310-314.
[330] Heraclides Ponticus proposed a daily Earth rotation in about 350 BC.
[331] Eternal Fascinations with the End: Why We're Suckers for Stories of Our Own Demise, by Michael Moyer,
September 2010 Scientific American Magazine.
[332] Based on the Google Books Ngram Viewer.
[333] Before Kuhn, the word paradigm had a narrower meaning. It might refer to a particularly important example in a field.
[334] Ziauddin Sardar, as quoted in NY Times, June 3, 2007.
[335] As quoted by Brown, 2005.
[336] Polanyi, Personal Knowledge (University of Chicago, 1958).
[337] Science: The Origin of Relativity, Time magazine, Jan. 26, 1970.
[338] Quantum Generations: A History of Physics in the Twentieth Century, by Helge Kragh. Princeton University Press, 2002. Chap. 7, p.90.
[339] Charles Percy Snow from C.P. Snow, Variety of Men, Penguin Books, Harmondsworth, U.K. 1969, pp 85-86.

[340] E.g., Marc Lange gives this argument.
[341] Robert Rynasiewicz, The Construction of the Special Theory: Some Queries and Considerations. Einstein Studies, vol. 8: Einstein: The Formative Years, pp. 159-201. (2000)
[342] The Evolution of Physics (1938) by Einstein and Infeld.
[343] Essential relativity: special, general, and cosmological, by Wolfgang Rindler, rev 2nd ed, Springer-Verlag, 1979.
[344] Kevin Brown, An Anomaly in Translation, mathpages.com.
[345] Jeremy Bernstein, Cranks, Quarks and the Cosmos, p.18, 20.
[346] Poincare's philosophy: from conventionalism to phenomenology, by Elie Zahar, p.2.
[347] Popper, Replies to my critics, 1971.
[348] The first observation of starlight deflection was going to test Einstein's erroneous 1911 prediction. Luckily for Einstein, the weather was bad, and he was able to correct his error before the 1919 observation.
[349] The Grand Design. p. 21, 96, 100.
[350] Hawking, The Universe in a Nutshell, p.31. He later repudiated this opinion.
[351] Science and Hypothesis, 1902.
[352] The Cult of Statistical Significance: How the Standard Error Costs Us Jobs, Justice and Lives, by Stephen T. Ziliak and Deirdre N. McCloskey. University of Michigan Press, 2008.
[353] Kuhn, The Copernican Revolution: Planetary Astronomy in the Development of Western Thought, Harvard University Press, 1992, p.171.
[354] Kuhn, Structure, p.75
[355] Ibid., p.172.
[356] Kuhn says, on p.152, "Some of these reasons -- for example, the sun worship that helped make Kepler a Copernican -- lie outside the apparent sphere of science entirely." And on p.156, "the Copernican theory scarcely improved upon the predictions of planetary position made by Ptolemy."
[357] Alfred Wegener explained it and gave very strong evidence for it in a 1925 book, but few people believed him.
[358] Kuhn, Black-Body Theory and the Quantum Discontinuity, 1894-1912. University of Chicago Press, 1987.
[359] Kuhn's student Errol Morris details this in a NY Times blog essay on The Ashtray, March 6-10, 2011.
[360] Popular Lectures and Addresses (1891-1894, 3 volumes).
[361] Freeman J. Dyson, Clockwork Science. The New York Review of Books, vol. 50, num. 17. Nov. 6, 2003.
[362] Stephen Jay Gould, from The Book of Life, W.W. Norton and Co., New York, 2001
[363] July 30, 2010 radio ads.

[364] Schroedinger, Are there quantum jumps? Brit. J. Phil. Sci., Part I, Aug. 1952, Part II, Nov. 1952.
[365] Born wrote a rebuttal to Schroedinger, The Interpretation of Quantum Mechanics, 1953, and claimed to have Bohr, Heisenberg, Pauli, and others on his side.
[366] Alan Woods, What Really Killed the Dinosaurs?, 2004.
[367] Heisenberg, 1958.
[368] Poincaré's philosophy: from conventionalism to phenomenology, by Élie Zahar. Carus Publishing, 2001. Introduction, p.2.
[369] Lecture on Geometry and Experience, Jan. 27, 1921.
[370] Cerf, Dismissing renewed attempts to deny Einstein the discovery of special relativity, Am. J. Phys., Vol. 74, No. 9, September 2006.
[371] Weinberg, Gravitation and Cosmology, John Wiley & Sons, 1972. He later adopted the geometric view, and used it in his 2008 book Cosmology.
[372] Darrigol, The Genesis of the theory of relativity, Séminaire Poincaré 1: 1–22.
[373] As quoted by his son, Hans Henrik Bohr, in a 1967 essay.
[374] Professor John Wheeler of Princeton University. He coined the term "black hole" and co-authored a popular relativity textbook.
[375] In Science and Hypothesis, 1902, ch.12, Poincare said, "The day will perhaps come when physicists will no longer concern themselves with questions which are inaccessible to positive methods, and will leave them to the metaphysicians. That day has not yet come; man does not so easily resign himself to remaining for ever ignorant of the causes of things."
[376] Hawking-Mlodinow, p.7. On p.41 it says, "So which is true, the Ptolemaic or Copernican system? Although it is not uncommon for people to say that Copernicus proved Ptolemy wrong, that is not true."
[377] Kepler, Epitome of Copernican Astronomy, Book IV, before Part I, 1618. He wrote, "Since astronomy has two ends, to save the appearances and to contemplate the true form of the edifice of the world ..."
[378] The coordinates are contracted, but the visual view to an observer looks more like a rotation. This was explained by Roger Penrose and James Terrell in 1959.
[379] Poincare, 1902, chap. 12.
[380] Miller seems not to have noticed that Poincare discussed the principle of relativity being a convention in Poincare's essay Space and Time, published in Last Essays, 1913.
[381] Tribe, The Curvature of Constitutional Space: What Lawyers Can Learn from Modern Physics, 103 Harvard Law Rev 1 (Nov. 1989).
[382] Wilson, The New Freedom. Doubleday, 1913.
[383] The Sokal Hoax: The Sham That Shook the Academy, 1999. Fashionable Nonsense: Postmodern Intellectuals' Abuse of Science by Alan Sokal and

Jean Bricmont, 1998. Beyond the Hoax: Science, Philosophy and Culture by Alan Sokal, 2008.
[384] Letter to Physics Today, Jean Bricmont and Alan Sokal, April 7, 1999.
[385] Can science explain everything? Anything? By Steven Weinberg, in NY Review of Books. Republished in The Best American Science Writing, 2002, ed. Matt Ridley.
[386] Sokal's Hoax, by Steven Weinberg. The New York Review of Books, Volume XLIII, No. 13, pp 11-15, August 8, 1996.
[387] Interview with Paul Dirac in the early 1970s for broadcast as part of a CBC radio documentary series entitled Physics and Beyond, and later published in Glimpsing Reality: Ideas in Physics and the Link to Biology, It is not clear whether the Einsteinian constant is supposed to be the gravitational constant or the speed of light.
[388] Quantum Quackery, by Victor Stenger. Skeptical Inquirer, Volume 21.1, January-February 1997.
[389] ABC TV Nightline Face-Off, March 2010.
[390] The Grand Design, p.140.
[391] Dirac 1987, p. 196.
[392] Raphael Bousso, Ben Freivogel, Stefan Leichenauer, Vladimir Rosenhaus, Eternal inflation predicts that time will end, arXiv:1009.4698v1, 2010.
[393] Josephson, String Theory, Universal Mind, and the Paranormal, 2003. He shared the 1973 Nobel Prize for physics for his work on superconductivity.
[394] Daryl J. Bem, Feeling the Future: Experimental Evidence for Anomalous Retroactive Influences on Cognition and Affect, Journal of Personality and Social Psychology, 2011.
[395] Susskind, Lectures on new revolutions in particle physics, Jan. 2010.
[396] As quoted in Superconductivity's Smorgasbord Of Insights: A Movable Feast, AAAS Science, 332 p.192, Apr. 8, 2011.
[397] Merriam-Webster's Medical Dictionary, Einstein.
[398] Garrett Lisi, An Exceptionally Simple Theory of Everything, 2007.
[399] String theory usually assumes that the Ricci tensor is zero, as in Grossmann's 1913 equations for empty space.
[400] Michio Kaku, blog posting.
[401] Colbert Report, Comedy TV channel, July 18, 2010.
[402] See "What about God?", chap. 11 in Dreams of a Final Theory, by Steven Weinberg, 1992.
[403] A Scientist Takes On Gravity, NY Times, July 13, 2010.
[404] As reported by Jo Marchant on a New Scientist blog, 18:10 3 August 2010.
[405] House Subcommittee on Energy and Environment of the Committee on Science and Technology, Sept. 30, 2009.
[406] Quoted on p.438 in The Strangest Man: The Hidden Life of Paul Dirac, Mystic of the Atom, by Graham Farmela. Basic Books, 2009.

[407] Quoted in String theory: Hanging on by a thread?, USA Today, Sept. 18, 2006.
[408] Twistor theory was invented by Penrose in connection with general relativity. It treats spinors as being more fundamental than spacetime.
[409] PBS Nova interview, 2003, S. Weinberg.
[410] Some estimates put the number at 10 to the power 500, more than the number of electrons in the known universe.
[411] June 2010 Scientific American magazine.
[412] Brian Greene and others said so in a 2010 panel discussion.
[413] As quoted on Amazon.com.
[414] The Grand Design, by Stephen Hawking, Leonard Mlodinow. 2010, p.166, 181.
[415] Many Kinds of Universes, and None Require God, by Dwight Garner, NY Times book review of The Grand Design, September 7, 2010.
[416] Hawking, Mlodinow, p.143.
[417] For example, in a July 31, 2006 KQED interview, Susskind said that Woit "failed as a physicist … he now is taking out his revenge". Woit got his PhD in physics, but is now a mathematician. Susskind did not address the substance of Woit's book.
[418] E.g., see string theorist Andy Strominger gave a report card at a 2010 Harvard colloquium.
[419] Woit, Not Even Wrong. Basic Books, 2006. Chap. 14, p.203.
[420] Letter to New Scientist, 08 December 2010, and ESI Special Topics, April 2002
[421] Posted on The String Coffee Table Blog, August 19, 2006.
[422] String theory: The fightback, by Amanda Gefter, New Scientist, 11 July 2007.
[423] Wilczek, Quantum Field Theory, arXiv:hep-th/9803075v2, 1998.
[424] Friedan, A tentative theory of large distance physics. arXiv:hep-th/0204131v1, 2002. It appears that the phrase "as it stands" was added to please an editor.
[425] Hawking, A Brief History of Time, Bantam Press 1988.
[426] The Grand Design, co-authored with Leonard Mlodinow, (Bantam Press 2010).
[427] Dreams of a final theory, Steven Weinberg, 1992. Vintage Books.
[428] Michael J. Disney, American Scientist magazine, Sept-Oct 2007.
[429] NY Times, January 15, 2008.
[430] Cover story, Scientific American, Jan. 2010.
[431] The Hidden Reality: Parallel Universes and the Deep Laws of the Cosmos, by Brian Greene, 2010.
[432] Structure and Evolution of the Stars, by Martin Schwarzschild, 1958.
[433] Alejandro Jenkins and Gilad Perez, Scieintific American, Jan. 2010.

[434] Susskind, p.9-10.
[435] Ibid., p.263.
[436] The time for a black hole to evaporate is much longer than a trillion years, and more like 10 to the power 68 years.
[437] E.g., see Einstein's Dream, a July 1991 Hawking lecture in Tokyo, published in Black Holes and Baby Universes and Other Essays, 1993.
[438] September 25, 1918 Einstein letter responding to 1914 book by Eduard Study, (CPAE 8, Doc. 624)
[439] Pauli credited Voigt, Lorentz, and Poincare in an encyclopedia article, and in his own book, Theory of Relativity. He also said that Einstein "in a way completed the formulation". Pauli, The Theory of Relativity and Science, Helvetica Physica Acta, Supp iv, p.282 286 (1956)
[440] In Born's book, Einstein's Theory of Relativity, published in 1924 and 1962.
[441] History of the Theories of Aether and Electricity, Vol. II (Glasgow, London: Nelson), 1953.
[442] Hawking, A Brief History of Time, 1988.
[443] The Rediscovery of Gravity, essay in It Must Be Beautiful: Great Equations of Modern Science, edited by Graham Farmelo. Granta Books, 2002. The Road to Reality, Roger Penrose. Random House, 2004.
[444] The Lightness of Being: Mass, Ether, and the Unification of Forces, by Frank Wilczek. Basic Books, 2008. It says, "The equations of electromagnetic relativity had, by 1905, been derived by Hendrik Lorentz and perfected by Henri Poincaré", p.79.
[445] Lecture at Caius College in May 1992, published in Black Holes and Baby Universes and Other Essays, 1993.
[446] As quoted in On Einstein's Weltbild, by Gerald Holton, republished in Albert Einstein's theory of general relativity, edited by Gerald Tauber. Crown Publishers, 1979.
[447] Einstein, The problem of space, ether, and the field in physics, Mein Weltbild, Amsterdam: Querido Verlag, I934. Republished in Ideas and Opinions, Crown, 1954.
[448] Einstein, NY Times, Nov. 9, 1930, quoted in 1955 obituary.
[449] Born letter to Einstein, 1944. Born's word for rubbish was *abfall.*
[450] David Gross, Einstein and the search for unification, 2005.

8. Timeline

500 BC Pythagoras
400 BC Plato, Aristotle
300 BC Aristarchus has heliocentric model
0150 Ptolemy, Almagest
1543 Copernicus, De Revolutionibus
1577 Tycho's comet
1582 Pope adds ten days to calendar
1604 Kepler's supernova
1609 Kepler, New Astronomy
1610 Galileo, Starry Messenger, telescope observations
1616 Galileo warned by the Inquisition
1633 Galileo punished
1676 Romer measures speed of light by watching Jupiter
1687 Newton, Principia
1725 Bradley measures aberration of starlight
1788 Lagrange formulates new mechanics
1796 Laplace conjectures black holes
1833 Hamilton formulates new mechanics
1805 Herschel discovers motion of the Sun
1838 Bessel measures stellar parallax
1851 Foucault's pendulum shows rotation of the Earth
1861 Maxwell finds equations for electromagnetism
1873 Maxwell, Treatise on Electricity and Magnetism
1877 Maxwell coins word "relativity"
1887 Michelson-Morley fails to detect motion of the Earth
1889 FitzGerald says contraction is a logical consequence

1992 Lorentz discovers local time, relativistic transformations
1995 Lorentz electron theory, relativity theorem
1999 Lorentz predicts relativistic mass
1900 Poincare, clock synchronization, mass-energy equivalence
1900 Planck proposes light quantum (photon)
1901 Kaufmann confirms relativistic mass
1902 Poincare, Science and Hypothesis
1904 Lorentz perfects theorem of corresponding states
1904 Poincare announces new mechanics from relativity principle
1905 Poincare discovers spacetime geometry, electromagnetic covariance
1905 Einstein publishes what is soon called the Lorentz-Einstein theory
1908 Minkowski popularizes relativity in spacetime
1913 Grossmann finds general relativity field equations
1913 Cartan discovers spinors
1913 Bohr model of atom
1915 Einstein improves Poincare's explanation of Mercury anomaly
1915 Hilbert, Schwarzschild find alternate formulations of general relativity
1916 Einstein, Relativity: The Special and General Theory
1918 Noether proves symmetries explain conservation laws
1919 Eddington eclipse makes Einstein famous
1919 Weyl unifies general relativity with gauge theory
1926 Heisenberg, Schroedinger, Born discover quantum mechanics
1927 Lemaitre discovers big bang
1928 Dirac finds relativistic quantum theory
1955 Einstein dies
1962 Kuhn, paradigm shift theory
1970 standard model of particle physics
1984 first string theory revolution
2006 Perelman wins prize for solving Poincare Conjecture

9. Index

CPSIA information can be obtained
at www.ICGtesting.com
Printed in the USA
FFOW01n1026110116
20318FF